Dario Davila Paredes

Structural analysis and floristic composition of a forest

Dario Davila Paredes

Structural analysis and floristic composition of a forest

of middle terrace, Puerto Almendra community, San Juan Bautista district, Loreto - Peru

ScienciaScripts

Imprint
Any brand names and product names mentioned in this book are subject to trademark, brand or patent protection and are trademarks or registered trademarks of their respective holders. The use of brand names, product names, common names, trade names, product descriptions etc. even without a particular marking in this work is in no way to be construed to mean that such names may be regarded as unrestricted in respect of trademark and brand protection legislation and could thus be used by anyone.

Cover image: www.ingimage.com

This book is a translation from the original published under ISBN 978-613-9-41080-4.

Publisher:
Sciencia Scripts
is a trademark of
Dodo Books Indian Ocean Ltd. and OmniScriptum S.R.L publishing group

120 High Road, East Finchley, London, N2 9ED, United Kingdom
Str. Armeneasca 28/1, office 1, Chisinau MD-2012, Republic of Moldova, Europe
Printed at: see last page
ISBN: 978-620-8-19708-7

STRUCTURAL ANALYSIS AND FLORISTIC COMPOSITION OF A MID-TERRACE FOREST, PUERTO ALMENDRA COMMUNITY, SAN JUAN BAUTISTA DISTRICT, LORETO - PERÚ

Dario Davila Paredes[1] , Génesis M. Davila Zambrano[2] , Rodil Tello Espinoza[3] , Jorge Mera Ramirez[4] , Roger Ruiz Pardes[5] , Ronald Burga Alvarado[3] , Elmer S. Saavedra Viteri[6] , Gladis Cardenas Cardenas Cardenas[8] , Richard Huaranca Acostupa[7] , Diego D. Davila Zambrano[7] , Pedro A. Angulo Ruiz[3] , Luis Lopez Vinatea[8]

1.   Amazon Herbarium, Centro de Investigacion Recursos Naturales (CIRNA), Universidad Nacional de la Amazonia Peruana (UNAP).

2.   Bromatology and Human Nutrition, Faculty of Food Industries, National University of the Peruvian Amazon (UNAP).

3.   Faculty of Forestry Sciences, National University of the Peruvian Amazon (UNAP).

4.   Faculty of Economics and Business, Universidad Nacional de la Amazonia Peruana (UNAP).

5.   Faculty of Food Industries, National University of the Peruvian Amazon (UNAP).

6.   Faculty of Education Sciences and Humanities, National University of the Peruvian Amazon (UNAP).

7.   Faculty of Biological Sciences, National University of the Peruvian Amazon (UNAP).

8.   Faculty of Chemical Engineering, National University of the Peruvian Amazon (UNAP).

TABLE OF CONTENTS

SUMMARY

The main objective of this research work was to analyze the floristic structure and composition of forest species, Puerto Almendra community, with emphasis on abundance, frequency, dominance and importance value index; complemented with the floristic complexity, horizontal and vertical structure, of the evaluated forest species. In three (3) plots of 20 m x 100 m, a total of 152 trees were recorded, represented by 23 families, 43 genera and 64 species; the most representative families were: Fabaceae, Lauraceae, Sapotaceae, Euphorbiaceae and Annonaceae, which reported 95 trees and represented the
62.5 %. The absolute and relative abundance, referred to quantify forest species, the most representative family Fabaceae, 6 genera, 10 species and 29 trees, *Parkia velutina* Benoist, 17 trees, totaling 52 trees, represented 34.21% of the total. Absolute and relative frequency; referred to the presence of forest species, in 04 plots, 04 forest species were recorded (40%), with the *Parkia* genus, "pashaco", standing out. Dominance, absolute and relative, of 152 trees, the most important family was Fabaceae, *Parkia velutina* Benoist, "pashaco" (40%); they totaled 5.615417 m^2 , of basal area. The Importance Value Index (IVI), *Parkia velutina* Benoist, "pashaco", 17 trees, represented 8.7% of the total. Horizontal structure (Eh), referred to the diameter, ranges from 5x5 cm to 10 cm dap, 1.30 m. from the ground. Diameter class I, (10-14.9 cm, DBH), quantitatively, registered 65 trees (42.76%), emphasized Sapotaceae family, *Micropholis egenesis* (A. DC.) Pierre, "caimitillo". Vertical structure (Ev), referred to total height (5-8 m), the lower vertical structure, quantitatively, registered 18 trees (11.84%), highlighted Lauraceae family, *Ocotea aciphylla* (Nees) Mez, "cinnamon moena"; the middle vertical structure, (Evm), quantitatively, registered 108 trees (71.The upper vertical structure (Evs), quantitatively, registered 26 trees (17.11%), with the Fabaceae family, *Parkia velutina* Benoist, "pashaco"; the Floristic Complexity was 1/3, that is, for each species, there were 3 trees.

Key words: Abundance; Frequency; Dominance; Importance Value Index; Floristic Complexity; Horizontal and Vertical Structure.

ABSTRACT

The main objective of this research work was to analyze the structure and floristic composition of forest species, Puerto Almendra community, with emphasis on abundance, frequency, dominance and importance value index; complemented with the floristic complexity, horizontal and vertical structure, of the forest species evaluated. On three (3) plots, 20 m x 100 m, a total of 152 trees were recorded, represented by 23 families, 43 genera and 64 species; The most representative families were: Fabaceae, Lauraceae, Sapotaceae, Euphorbiaceae and Annonaceae, reporting 95 trees and representing 62.5%. The absolute and relative abundance, referred to quantifying forest species, the most representative family Fabaceae, 6 genera, 10 species and 29 trees, highlighted Parkia velutina Benoist, 17 trees; They added a total of 52 trees, representing 34.21% of the total. The absolute and relative frequency; Regarding the presence of forest species, in 04 plots, 04 forest species were recorded (40%), the Parkia genus, "pashaco", stood out. Dominance, absolute and relative, of 152 trees, the most important family was Fabaceae, highlighted Parkia velutina Benoist., "pashaco" (40%); They added up to a total of 5.615417 m2 of basal area. The Importance Value Index (IVI), highlighted Parkia velutina Benoist., "pashaco", 17 trees, represented 8.7% of the total. Horizontal structure (Eh), referring to the diameter, ranges from 5x5 cm to 10 cm dbh, 1.30 m. ground. Diametric class I, (10-14.9 cm, DBH), quantitatively, registered 65 trees (42.76%), highlighted by the Sapotaceae family, Micropholis egenesis (A. DC.) Pierre, "caimitillo". Vertical structure (Ev), referred to total height (5-8 m), the lower vertical structure, quantitatively, recorded 18 trees (11.84%), highlighted Lauraceae family, Ocotea aciphylla (Nees) Mez, "canela moena"; the medium vertical structure, (Evm), quantitatively, recorded 108 trees (71.05%), highlighted the Lecythidaceae family, Eschweilera coriaceae (A. DC.) S. A. Mori, "machimango"; the upper vertical structure, (Evs), quantitatively, recorded 26 trees (17.11%), highlighted Fabaceae family, Parkia velutina Benoist, "pashaco"; Floristic Complexity was 1/3, that is, for each species, there were 3 trees.

Keywords: Abundance; Frequency; Dominance, Importance Value Index; Floristic complexity; Horizontal Structure and Upright structure.

INTRODUCTION

Peru, a forest country, with 60% of forests; it occupies the tenth place in forest cover in the world, second in Latin America, after Brazil[1] , represents to the Peruvian Amazon basin, 956 751 Km^2 , (74.44 %) of the national territory and 13.02 % of the total Amazon basin[2] . The forests of the Peruvian Amazon are richer in species diversity than any other tropical forest on the planet, 50% corresponds to the Amazon plain or plain; of 17 144 species registered in the catalog of Angiosperms and Gymnosperms of Peru, 43% of the plants are Amazonian[5] ; the floristic composition represents 6 237 species, distributed in 1406 genera and 182 families, constituting 36.3% of the phanerogamous flora of Peru[7] , it is possible to find up to 300 species of plants in one hectare[6] , almost 3000 species in three reserved zones of Iquitos .[8]

Forest inventories are valuable instruments to record information and show the situational status of the forest, from a quantitative and qualitative point of view[3] . The natural forests located in the community of Puerto Almendra have a little known natural vegetation of medium terrace, not yet documented and the role of the species of this type of forest is unknown; however, they are currently exposed to various anthropogenic activities, the main threat being deforestation[4] , for housing and timber sales, which causes the loss of an ecosystem with valuable natural habitats for the Amazonian flora and fauna, especially for the inhabitants of this place.

Evaluating and quantitatively establishing their structural characteristics allows us to determine the possibilities of use, in production and conservation, as well as the scientific, technical, economic and ecological interest, to guide a successful sustainable management of these ecosystems and develop the Loreto Region.

The horizontal and vertical structure evaluates the behavior of each species of a forest type, through indexes that express species occurrence and ecological importance within the ecosystem, through abundance, frequency and dominance, whose relative sum generates the IVI .[9]

The objective of this work is to quantitatively analyze the most important structural characteristics and floristic composition of a medium terrace forest, located in the community of Puerto Almendra, San Juan Bautista District, Maynas Province, Loreto Region. Due to its geographic location, it is considered an ecosystem of special interest for the Peruvian Amazon.

I. THEORETICAL FRAMEWORK

1.1. BACKGROUND

Studies of floristic composition and diversity of tropical forests are important and essential to understand the structure and dynamics of vegetation [59, 60, 61, 62, 63].

Numerous forest inventories carried out in the Peruvian Amazon conclude that the families Fabaceae, Lauraceae, Annonaceae, Rubiaceae, Moraceae, Myristicaceae, Sapotaceae, Meliaceae, Arecaceae and Euphorbiaceae contribute 52% of the species richness of the low altitude forests of the Peruvian Amazon .[58]

Studies conducted on the flora in 3 reserves of Iquitos (Allpahuayo-Mishana, Yanamono and Sucusari), showed that the terra firme forests are richer in species than the floodplain forests. Of the species recorded, 74.6% occur only on dry land, 16.2% grow in floodplains and 9.2% of the species grow in both cases .[59]

In 2012, a study was conducted on a mid-terrace forest adjacent to the Arboretum "El huayo" of the Forestry Research and Teaching Center, Puerto Almendra; the floristic composition, horizontal and vertical structure of trees with - 10 cm DBH were determined; 1,159 trees were recorded, distributed in 38 botanical families, 92 genera and 120 species; the most important botanical family was Euphorbiaceae with 10 species and 222 individuals (54%, IVI); ecologically the most important species was *Alchomea triplinervia*, "zancudo caspi" (Euphorbiaceae), with 20.2% of IVI; *Virola flexuosa* "cumala caupuri" (Myristicaceae), with 18.59%; *Protium* sp. "copal" (Burseraceae), with 17.4%; *Micrandra spruceana* "shiringa masha" (Euphorbiaceae), with 16.33% .[97]

In 2016, in a forest inventory, in primary forest influenced by the Iquitos-Nauta highway, the forest composition was evaluated and analyzed in 5 hectares, trees with - 10 cm DBH, at 1.30 m from the ground, 1 614 trees belonging to 161 different species were recorded. The species with the highest abundance was "copal" (*Protium* sp.), 80, (4.96%), as well as the Importance Value Index[10] . The vertical structure of the vegetation was considered in three strata: lower, middle and upper. The middle stratum was comparatively the most uniform, most closed and with the highest number of trees. The lower stratum was comparatively uniform, less closed with fewer trees. The upper stratum was mostly sparse, the canopy of the trees separated to open, few species in this canopy apparently of great horizontal spread (dominance)[10] . The horizontal structure of the inventoried trees was distributed in 19 diameter classes. The highest percentage of trees, (70.82%), were distributed in the diameter classes of 10 < dbh < 25 cm and represented 27.94% of the basal area .[10]

In 2018, the evaluation of the floristic composition and horizontal structure of the arboreal vegetation of the Arboretum "El huayo" was carried out; it was determined the intensity of mixture, index value of importance of species and families; in 4 permanent sampling plots of 1.22 ha. each; all individuals were recorded with - 5 cm, dap; 521 forest species distributed in 183 genera and 53 families were found. The most ecologically important family was Lecythidaceae, because it adapted better to the site and had a higher IVI .[12]

In 2019, a slightly dissected low hill forest was evaluated in 13 sampling plots of 20 x 250 m, Iquitos-Nauta road, Maynas province, department of Loreto; the floristic composition, horizontal structure and diversity of this forest was determined; it reported 709 individuals distributed in 105 species, 86 genera and 33 botanical families. Twenty (20) species contribute the greatest ecological weight, *Tachigali sp.* with 14.72%, *Eschweilera decolorans* with 12.53% and *Virola elongata* with 11.72% .[13]

During 2020, a forest inventory was conducted in a terrace forest, Madre de Dios region, the floristic composition, structure and tree diversity of an Amazonian forest was determined, in 2 rectangular plots of 20 m x 500 m each, individuals were identified, with - 10 cm dap; 4 429 trees, 254 species, 165 genera and 53 families were recorded. The most ecologically important species were: *Tetragastris altissima* (Aubl.) Swart, *Iriartea deltoidea* Ruiz & Pav, *Euterpe precatoria* Mart., abundant species due to the opening of the canopy, which increased their dominance; while *Ocotea bofo* Kunth, *Bertholletia excelsa* Bonpl, *Eschweilera coriácea* (DC.) S.A. Mori, with low abundance, due to changes in the floristic composition, after the use of wood .[14]

In 2021, a study was conducted in nine locations, Loreto Region, to compare patterns of composition, structure and floristic diversity in different types of forests, in plots 20 m x 50 m (2.5 ha.). The families with the highest abundance were: Fabaceae, Euphorbiaceae, Myristicaceae, Annonaceae, Lecythidaceae, Moraceae, Violaceae, Burseraceae, Rubiaceae and Sapotaceae. According to the IVF, the specific richness was 903 species, with 3421 individuals. The average basal area of 76.92 m /ha .[2(15)]

1.2. THEORETICAL BASES

1.2.1 Structure of forest

Ecologically, the structure of a forest is the arboreal component that is in direct relation with the forces of the environment, mainly with the climate, physiography, soil, elements favorable for the growth and development of other forms of life, It interrelates with the soil through the contribution of organic material and nutrients, which reduces and buffers the climatic effects on the soil and its physiography, constituted by the arboreal vegetation, comprised of woody plants capable of exceeding 10 cm in diameter, an indicator of the establishment in the main forest canopy .[16]

The structure of the forest is given by the abundance, distribution and dominance of the species that make it up, with respect to the total forest mass. The sum of these relative parameters indicates the index value of ecological importance and its dominance in the forest .[17]

Most species have a low horizontal distribution, with irregular to scarce occupation within the forest ("occasional" species). Of the floristic group, the horizontally well distributed species ("frequent" species), usually with high abundance and greater dominance, play an important role in the constitution of the forest .[18,19,20]

The most commonly used parameters to measure the structure of a forest are stem density and basal area[23] . Vertically we can distinguish three tree strata: upper stratum, middle stratum and lower stratum.
[21,19,22,20].

1.2.2 Composition floristic

The characteristics of tropical forests are determined by environmental factors, geographical position, climate, soil and topography, as well as the dynamics of the forest and the ecology of its species[26] . In addition, the floristic composition of tropical forests is constantly changing from one place to another; focusing on the diversity of species within an ecosystem, it is necessary to elaborate a table containing names of the identified species to describe them adequately .[23]

The high diversity of Amazonian forests compared to the rest of the Neotropics, is due to biodiversity by specialization to different habitats[25] . Floristic inventories in three forest reserves near Iquitos (Yanamono, Reserva Turística Explorama, Sucusari), report 2,748 species, 876 genera and 165 families; world record of local diversity in the Amazonian forests.
1.0 ha, 300 different species and 600 individual plants were recorded, with DBH greater than 10 cm[26] . The second richest plot in species diversity in the world is located in Mishana, Nanay River, with 289 species.
[26].

Studies of floristic composition and vegetation structure are fundamental to plan and develop techniques for the conservation and sustainable use of ecosystems and their components, their knowledge provides a broad vision of the biological mechanisms that occur there,[27] , essential to understand the dynamics of forests and the changes induced by human activity .[28]

The above ratifies the value of floristic inventories to answer questions such as: how much diversity exists, floristic composition, diverse families, abundant species, dominant species, structure, strata, endemism, and so on.

1.2.3 Complexity floristic

Floristic complexity is measured by the mixing coefficient which reflects the proportion of species abundance, referring to the degree of intensity of species mixing in a given area[30] . In Amazonian forests, the mixing coefficient varies between 1:3 and 1:4[31] . The same author, in Colombia, identified 1:7 as the approximate mixing coefficient for that area. The mixing coefficient for the primary forest in Tambo Quemado-Bolivia was 1:6 .[32]

Floristic complexity is one of the factors limiting the profitable management of forests[33] . However, this situation can be overcome by grouping species according to ecological temperament[34] , and commercial value[35] , reducing floristic complexity through silvicultural treatments.

1.2.4 Measurement of the structure

The first attempts to describe the structure of forest communities used parametric models, in terms of proportional abundance of each species, which describe the graphical relationship between the value of ecological importance of the species in terms of the sequential arrangement, by intervals of the species, from greater to lesser importance; these models differ in terms of the biological and statistical interpretations that assume the data .[36,37]

1.2.4.1   Abundance

The absolute and relative abundance of species is the total number of trees belonging to a given species[38,23,11] . Participation of the number of trees per species as a percentage of the total number of trees inventoried in the plots.

1.2.4.2   Frequency

The frequency, absolute and relative, measures the regularity and horizontal distribution of each species in the forest, that is, its mean distribution. To determine it, the sample (plot) is divided into a convenient number of subplots of equal size. It controls the presence or absence of the species in each subplot[38,23,11] . The absolute frequency of a species is expressed as a percentage of the subplots in which the species occurs (total number of subplots = 100%). Relative frequencies are calculated based on the sum total of the absolute frequencies of a sampling which is considered equal to 100%.

1.2.4.3   Dominance

The absolute and relative dominance, or horizontal spread of species, "section determined on the ground surface by the horizontal projection beam of the plant body", i.e. the vertical projection of the crown of each tree[38,31]. To overcome this difficulty, it is proposed to use the basal area, expressed in m^2 [38], of the trees instead of the crown projection[39,11], thus expressing tree dominance is almost inapplicable in most tropical forests because of their complex horizontal and vertical structure.

1.2.4.4   Importance Value Index, (IVI)

The study of abundance, frequency and dominance reveal essential aspects of the floristic composition of the forest, but they are always partial approaches that in isolation provide the required information on the floristic structure of the vegetation as a whole and as such. A method to integrate abundance, frequency and dominance, consists in the calculation of the proposed Importance Value Index, thus gathering the most important species of a forest zone; the method to integrate these three aspects is the IVI, obtained by adding for each species its abundance, frequency and relative dominance, proposed by Curtis and Mc-Intosh Mc-Intosh .[38]

1.2.4.5   Horizontal floristic structure, (Eh)

The horizontal structure of a population or forest can be described by the distribution of the number of trees per diameter class, the basal area being an index of the degree of development or as an indicator of the competence of a forest, but can also reflect the degree of intervention that has occurred .[23]

From a silvicultural point of view, the most important measure of horizontal organization is basal area, which can be used as an index of the degree of development or as an indicator of the competence of a forest, but can also reflect the degree of intervention that has occurred[23]. Each diameter class indicates the diameter range at which the basal area of the plants that determine the horizontal structure of a stand has been calculated.

1.2.4.6   Vertical floristic structure, (Ev)

Vertical floristic structure informs about the floristic composition of the different strata of the forest vertically and the role played by the different species in each of them[23]. The vertical structure of the five
(05) plots is determined by the distribution of trees along the top of their profile .[23]

In the tropical rainforest it is possible to distinguish:

a)   The upper stratum comprising the trees whose crowns form the highest canopy of the forest.
b)   The middle stratum comprising the trees whose crowns are below the highest

canopy.

c)   Lower stratum: tree canopies are located in the lower half of the canopy.

To determine the sociological position (canopy) of the trees, the following scale was used: upper stratum, middle stratum and lower stratum .[19]

1.2.4.7   Floristic complexity, (cf)

The floristic complexity expresses the intensity of mixture of each species in the forest, i.e. the number of trees per species, which expresses the intensity of mixture of all the evaluated plots[23] , because in a plant community composed of trees, bushes and sometimes shrubs, which form a few overlapping strata .[41]

1.2.4.8   Diametric distribution of species

The horizontal structure of a population or forest can be described by the distribution of the number of trees per diameter class and from a silvicultural point of view, the most important measure of horizontal organization is the basal area, which can be used as an index of the degree of development or as an indicator of the competence of a forest, but can also reflect the degree of intervention that has occurred .[23]

In order to make comparisons with other similar studies and according to general standards on standardization[75] , a class interval was set for the study - at 10 cm corresponding to the DBH .[23]

Table N° 01: Diameter Class and Range [23]

Diameter classRange	(cm) I
10-14.9	
II15-19 .9	
III20-24 .9	
IV25-29 .9	

1.2.4.9   Distribution by class of height

Based on structural and floristic studies of an Amazonian forest, the number of trees per height class was determined using the formula developed for the distribution of the number of trees per diameter class from 5 meters in height, vertical structure of 5 and 8 meters, medium vertical structure of 9 and 17 and upper vertical structure of 18 and 25 meters in total height, respectively .[76,74]

1.3. DEFINITIONS OF BASIC TERMS

11

Floristic composition: describes in terms of plant families, genera and species the individuals recorded during data collection .[42]

Medium terrace forest: forest community that develops on medium terraces, occasionally flooded, and can go several years without being reached by water .[43]

Forest structure: distribution of species in sizes and ages of a forest; with vertical (height) and horizontal (diameter) growth and tree succession; ecologically, it is in direct relation to environmental forces, mainly climate, physiography and soil .[44]

Structure: It is an element or set of elements joined together, with the purpose of supporting different types of stresses .[47]

Vertical structure: vertical organization of a forest, with the distribution of leaf masses in the vertical plane .[45]

Horizontal structure: Distribution of individual features, within the space of a wooded area, related to size, location and types of life form .[46]

Abundance: total number of trees belonging to a given species .[23]

Frequency: horizontal distribution of each species, represented by the absolute frequency, number of samples where a species is found and relative frequency, total frequency of all species .[49]

Dominance: measures the productive potential of the forest and is a useful parameter for determining the quality of the site[9] . It is represented by absolute dominance, the sum of the basal area of individuals belonging to a species, and relative dominance, a value expressed as a percentage of the total sum of absolute dominance .[49]

Importance Value Index (IVI): sum for each species of its abundance, frequency and dominance .[38,24]

Total height: vertical distance between the base and the apex of the tree .[51]

Diameter at breast height (DBH): diameter of a tree measured at a reference point, usually at 1.3 m from the ground .[52]

Basal area: cross-sectional area at chest height .[53]

Floristic complexity: Diversity index that expresses the variety of a forest, i.e. a ratio between the number of species and the number of individuals .[54]

II.   VARIABLES AND HYPOTHESES

2.1. Formulation of the hypothesis

2.1.1 General

Structural analysis and floristic composition of a medium terrace forest, Puerto Almendra community; allows estimating the importance and ecological potential of the area.

2.1.2 Specific:

-       **Alternatively**, the structural analysis and floristic composition of a medium terrace forest, Puerto Almendra community, allows estimating the importance and ecological potential of forest species, improving basic guidelines for an economic, integral and sustainable management of the tropical forest.

-       **Null**: The structural analysis and floristic composition of a medium terrace forest, Puerto Almendra community; does not allow estimating the importance and ecological potentialities with forest species, concluding with basic guidelines for an economic, integral and sustainable management of the tropical forest.

2.2. Operationalization of variables

2.2.1 Variables and their operationalization

The variables of interest are the structural analysis and floristic composition of a mid-terrace forest, in plots and subplots of forest species; being quantitative and continuous. The characterization variables are the type of medium terrace forest, being quantitative and nominal.

The following table shows the operationalization of the variables of the research work.

Table N° 02: Variables and their operationalization

Variable	Conceptual definition	Operational definition	Indicator	Items	Instrument
Variable of interest					
Structure and composition floristics of a terrace forest	Unknown natural vegetation, loss of habitats natural flora and fauna	Analysis statistician I.V.I.	- Forest type - Structure horizontal of species - Structure	- Abundance (%) - Frequency (%) - Dominance (%) - Total height (m) - Diameter (cm) - Coefficient of	Inventory data, calculation of horizontal structure and vertical of the forest, composition

media.	by anthropogenic activities.		of vertical species	mix.	floristic families, genera and species, format.
Characterization variables					
Forest of terrace half	Forest with flat relief, water level height 5-10 m, pending 0 y 8 %, vertical and horizontal vegetation structure defined.	The number of families, genera and species of the floristic composition of a terrace forest media.	- Structure and composition floristic	- Number of trees - Number of species - Number of genres - Number of families	Format of species registration and inventory forestry existing in a mid-terrace forest.

III. METHODOLOGY

3.1. Type and design of research

Non-experimental, quantitative, explanatory, descriptive, cross-sectional and comparative type, established by the analysis of structural characteristics and floristic composition of forest species, of a medium terrace forest; located in the community of Puerto Almendra, San Juan Bautista, Loreto, Peru.

3.2. Population and sample

The population is conformed by all the trees of the forest species in one hectare (10 000 m^2), whose coordinates were P1: 3°50'26.00 "S, 73°22'11.34 "W, 112 masl; P2: 3°50'43.66 "S, 73°22'13.53 "W, 114 masl; P3: 3°50'43.66 "S, 73°22'13.53 "W, 114 masl; P2: 3°50'43.66 "S, 73°22'13.53 "W, 114 masl; P3:
3°50'40.00 "S, 73°22'15.00 "W, 106 masl; P4: 3°50'38.00 "S, 73°22'12.00 "W, 106 meters above sea level, individuals with -10 cm DBH; sample represented by three plots of 20 m x 100 m, divided in nine subplots of 20 m x 20 m, subdivided in 10 m x 10 m, 5 m x 5 m and 2 m x 10 m, and 2 m x 10 m, subdivided in 10 m x 10 m, 5 m x 5 m and 2 m x 10 m.
2 m (0.6 ha.), Whittaker's method, widely used in Peruvian Amazonian

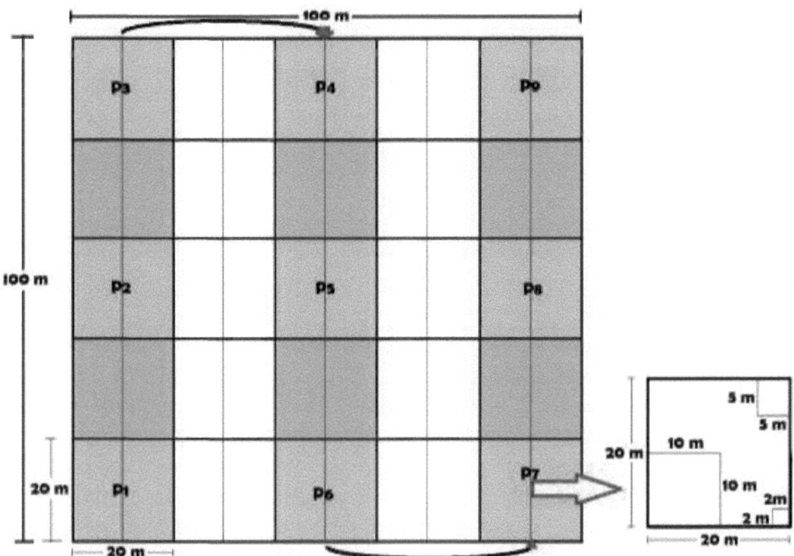

Illustration N° 01: Design of plots and subplots 20 m x 100 m (0.6 has.)

3.3. data collection procedures

It was obtained from primary sources: data or values recorded in field format: number of trees per species, plots and subplots, diameter and height of each tree, measured at 1.30 m from the ground (DBH); secondary sources include documents related to the Importance Value Index, complemented with horizontal and vertical structure, floristic complexity.

3.4. Techniques and instruments

3.4.1 General description of the area of study

The community of Puerto Almendra, located on the right bank of the Nanay River, a left tributary of the Amazon River, 22 km. from Iquitos in a south-south direction.

West[67] . Geographically located at coordinates 3°49'48" S and 73°25'12" W; altitude approx. 122 masl; limits to the North, with fiscal land that separates the right bank of the Nanay River, Almendra and Caño de la misma; East, Zungaro Cocha hamlet, Cocha of the same name and vacant land; West, with land of the hamlet of Nina Rumi, Llanchama Cocha and vacant land[68] . The climate is tropical, warm and rainy, with high atmospheric humidity between 80% and 100%. Average annual temperature between 24°C and 26°C, and maximum temperatures between 33°C and 36°C. Precipitation ranges from 2500 to 3000 mm. Soil, includes dry land, non-flooded land, developed from older alluvial deposits, including sediments deposited during the Tertiary and Quaternary, as well as alluvial plain, land subject to periodic flooding, due to recent deposits of the Amazon River and its tributaries, Itaya, Nanay and Tamshiyacu[57] . Ecologically, it belongs to the tropical rainforest type (bhT)[69] . To get to the community of Puerto Almendra, we use two means of communication, taking the city of Iquitos as a reference point: by paved road and by

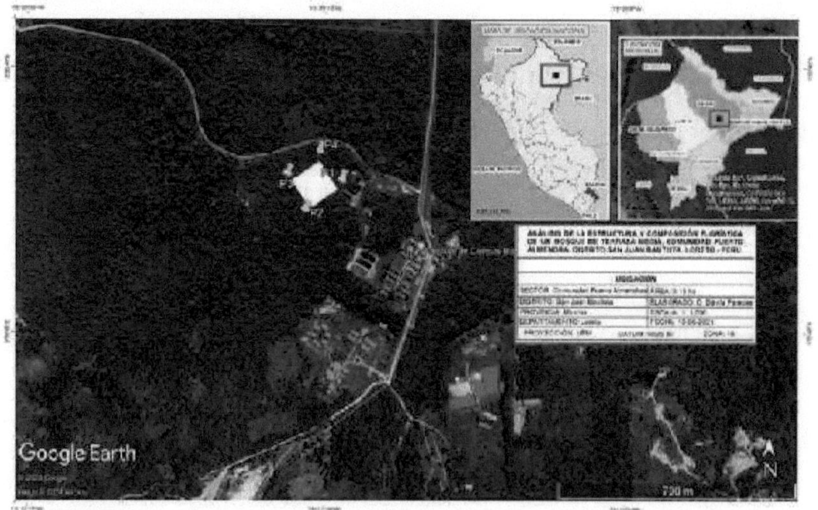

river, Nanay River.
Illustration N° 02: Location of the study area.

▶   **Materials**

Field: Machetes, winch, suunto compass, phosphorescent tape, indelible marker, raffia, field notebook, clinometer, telescopic scissors, crow's foot ascender, GPS, pulls.

► **From cabinet**: PRODESK computer, Intel Core 7, vPro, Float scientific calculator, stationery and specialized bibliography.

3.4.2 Methodology

Illustration N° 03: Exploratory visit and location of the study area

The work began with the recording of field data, in a format prepared by the author: common name, scientific name, DBH, total height, (ANNEX N° 01), all species located within the medium terrace forest were considered, 100% forest inventory technique. The technical instruments used were previously calibrated and accepted [66,67,17].

Illustration N° 04: Georeferencing, orientation and transect opening

To delimit plots, a SUUNTO compass, a 50-meter string and a 50-meter winch were used to draw three rectangular areas of 20 m x 100 m, divided in five (05) subplots (0.6 ha). Numbering and marking: each tree was numbered with a sequential code (1, 2, 3, n...), the marking was with phosphorescent tape, better view of the species within the forest. Diameter measurement: with diametric tape, the DBH was recorded (- 10 to more cm), all individuals in the study area, at a height of 1.30 m from the ground. Height measurement: with a 4 m telescopic rod, the total height of each tree was recorded from the base of the trunk to the apex.

Illustration N° 05: Coding and pressing of botanical samples collected Three (03) samples were collected per species, fertile or sterile, preserved with alcohol 50:50, placed in polyethylene bags of 20 cm x 67 cm x 5 cm and transferred to the Amazonense herbarium, Natural Resources Research Center, National University of the Peruvian Amazon, for drying of the samples, they were placed in used newspaper, interspersed with cardboard and corrugated aluminum, tied in a pile, subjected to an electric resistance dryer, temperature of 60 °, for 24 hours and; Identification was by comparison with exssiccata deposited at AMAZ, identification keys, specialized bibliography and the support of the specialized personnel of the Amazonian Herbarium - AMAZ .[70,71]

3.5. Processing techniques and analysis of data

The data were recorded in a Microsoft Excel 2011 database.

The analysis and interpretation of data were quantitatively processed by means of descriptive statistics, presented in tables, graphs and illustrations of relative abundance ($A\%$), relative frequency ($F\%$), relative dominance ($D\%$), expressed through the Importance Value Index, IVI[39] and floristic complexity; complemented with vertical and horizontal structure of the representative forest species of the area, personalized with a diagram of the forest profile, showing the sociological position (canopy) of the trees evaluated .[40]

3.5.1 Parameters Calculated

► **Horizontal structure, (E_h)**

For the floristic composition of the plot under study, a vegetation table was prepared, estimating the ecological weight of the species within the forest, by determining the absolute and relative values of abundance, frequency and dominance, added for the index value of ecological importance of the families, species and basal area .[74]

Absolute Abundance, (A_a)

Expresses the total number of individuals of the species.

$$A = \textstyle\sum n_i$$

Where:
n: trees of the same species.

Relative abundance, ($A\%$)

Indicates the participation of individuals of each species in percentage.

$$A\% \; \frac{A}{\sum t} = 100$$

Where:
$A\%$ = total absolute abundance of inventoried trees.
$\sum t$ = total sum of trees of all species.

Absolute frequency, (F_a)

$$F_a \; \frac{N°sp}{N°tsp} = 100$$

Where:
$N°sp$: total number of subplots in which a given species occurs. $N°tsp$: total number of subplots in which the plot was divided.

Relative frequency ($F\%$)

The number of subplots where the species occurs was considered.

$$F\% \; \frac{F_a}{\sum tF} = 100$$

Where:
F_a = absolute frequency.
$\sum tF$ = total sum of absolute frequencies of all species.

Absolute dominance ($_{Da}$)

Total sum of basal areas of individuals of all species.

$$Da = {}^{TM}ab$$

Where:
Ab = basal area, in m^2 , of trees of the same species.

Relative dominance, ($_{D\%}$)

It is the value expressed as a percentage of absolute dominance.

$$D\% \ \frac{Da}{{}^{TM}tab} = 100$$

Where:
$_{Da}$ = absolute dominance.
^{TM}tab = sum total basal area of all species.

Importance Value Index, (IVI)

$$IVI = {}_{A\%} + {}_{F\%} + {}_{D\%} = A\% + F\% + D\%$$

Where:
$_{A\%}$ = relative abundance. $_{F\%}$ = relative frequency. $_{D\%}$ = relative dominance.

► **Vertical floristic structure, ($_{Ev}$) Sociological Position**
It was determined based on the projection of the total height class curve versus the number of individuals, with the following scale .[19]

Table N° 03: Sociological Position [19]

Lower Stratum (IS)	less than 1/3 of the maximum height
Middle stratum (MS)	between 1/3 and 2/3 of maximum height
Upper stratum (ES)	greater than 2/3 of the maximum height

This procedure means the simplified numerical value of the percentage of the number of trees corresponding to each stratum, in relation to the general total of trees in all strata of the forest .[77,78]

$$\text{Floor } X = \frac{N° \text{ trees on the floor} \times 100}{\text{Total number of trees}}$$

A diagram, profile of 3 plots, 20 m x 100 m, was elaborated using the method described by the author.

▶ **Floristic Complexity, (c_f)**

It was evaluated by the Mixing Quotient (MC), which indicates the intensity of forest mixing, expressed as the quotient between the number of species (N_{sp}) and the number of individuals (N_{ind})[54] . The following formula was used:

$$C_f = (N_{sp}) / (N\acute{a}rbs)$$

Where:
(N_{sp}) = number of species. ($N\acute{a}rbs$) = total number of trees.

▶ **Diametric distribution of species**

The horizontal structure of a population or forest can be described by the distribution of the number of trees per diameter class. From a silvicultural point of view, the most important measure of horizontal organization is the basal area, used as an index of the degree of development or indicator of the competence of a forest [23].

To calculate the basal area of each tree, we use the formula:

$$ab = 0.7844 \times (dap)^2$$

DBH = Diameter at breast height.

▶ **Distribution by height class**

It was determined by the number of trees per height class, calculated using the formula developed by the distribution of the number of trees per diameter class, from 5 m to more than total height, - 10 cm dap, at 1.30 meters from the ground [76,74].

3.6. Ethical aspects

The criteria and reliability of the work is considered auditability, comprising a set of activities of the researcher and advisor, with responsibility, ensuring that this work is developed within academic standards and a quantitative consensus of the research, protecting and respecting human conduct, directly and indirectly, as well as privacy, confidentiality and consent of all those who participated.

IV. RESULTS

4.1. Plot N° 1

4.1.1 Abundance, absolute and relative

A total of 40 trees were inventoried, belonging to 11 families, 16 genera and 21 different forest species.

The most abundant local trees were: *Ecclinusa lanceolata* (Mart. & Eichl.) Pierre, "caimitillo", 06 trees, (15.0% of the total); *Parkia velutina* Benoist, "pashaco", 6 trees, (15.0% of the total); *Sloanea guianensis* (Aubl.) Benth., "casha huayo", 04 trees, (10.0%). These 03 species, totaling 16 trees, represented 40.0% of the total recorded (Table N° 04 and Graph N° 01).

Table N° 04. Absolute and relative abundance, most representative forest species of the site.

Scientific Name	Common Name	ABUNDANCE		
			absol	relat (%)
Ecclinusa lanceolata	"caimitillo"	6		15
Parkia velutina	"pashaco"	6		15
Sloanea guianensis	"casha huayo"	4		10
	sub total		16	40

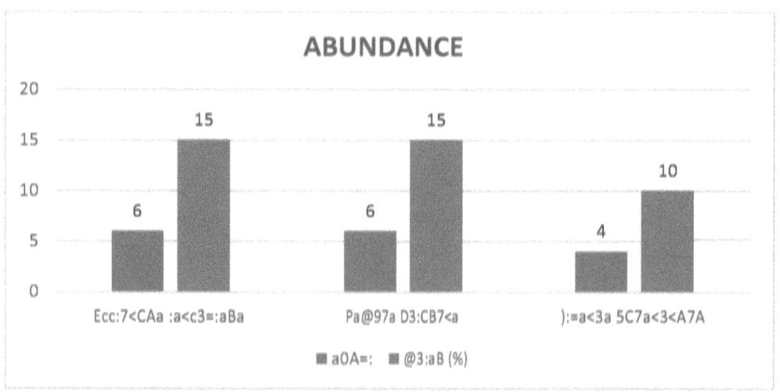

Gráfico N° 01. Abundancia absoluta y relativa, de especies forestales más representativas del lugar

4.1.2 Frequency, absolute and relative

Frequency was evaluated, dividing into 1 subplot.

The 21 local trees each had an absolute frequency value of 1 and a relative frequency value of 4.55%. (Table N° 05 and Graph N° 02).

Table N° 05. Absolute and relative frequency of the most representative forest species of the site.

Scientific Name	Common Name	FREQUENCY absol (N°)	relat (%)
Ecclinusa lanceolata	"caimitillo"	1	4.55
Parkia velutina	"pashaco"	1	4.55
Sloanea guianensis	"casha huayo"	1	4.55
sub total		3	13.64
Total		21	100

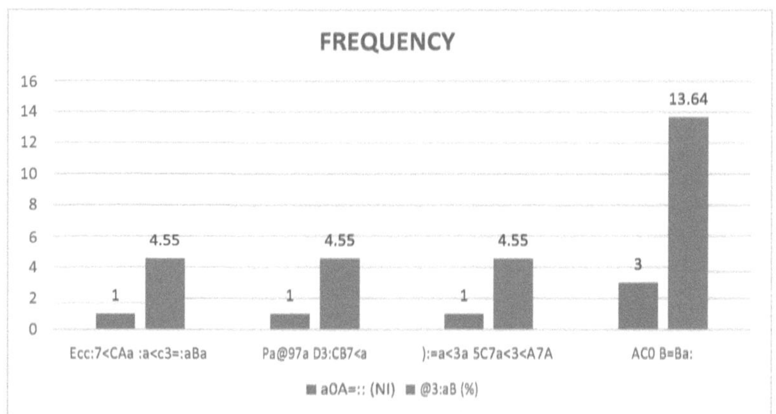

Absolute and relative frequency of the most representative forest species of the site.

4.1.3 Dominance, absolute and relative

Referred to the basal area of each of the 40 trees, belonging to 23 different local forest species.

The species with the highest absolute and relative dominance were *Ecclinusa lanceolata* (Mart. & Eichl.) Pierre; "caimitillo", with 0.25 m^2 , 16.93% of the total; followed by *Parkia velutina* Benoist; "pashaco", with 0.14 m^2 , (9.79%); *Sloanea guianensis* (Aubl.) Benth.; "casa huayo", 0.11 m^2 , (7.50%). These 03 local species totaled 1.46 m^2 basal area, representing 34.21% of the total (Table N° 06 and Graph N° 03).

Table N° 06. Absolute and relative dominance (m^2) of the most representative forest species of the site.

Scientific Name	Common Name	DOMINANCE Absol (m2)	Relat (%)
Ecclinusa lanceolata	"caimitillo"	0.25	16.93
Parkia velutina	"pashaco"	0.14	9.79
Sloanea guianensis	"casha huayo"	0.11	7.50
sub total		0.50	34.21
Total		1.46	100

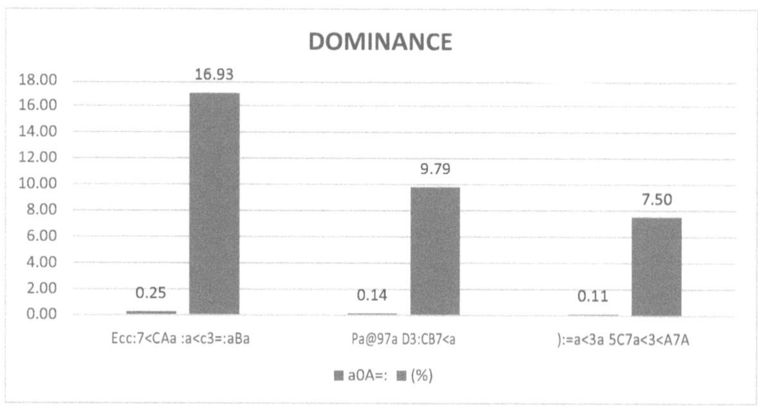

Graph N° 03. Absolute and relative dominance (m^2) of the most representative forest species of the site.

4.1.4 Importance Value Index, (IVI)

A total of 23 different forest species were calculated, grouping a total of 40 trees.

The species with the highest IVI were: *Ecclinusa lanceolata* (Mart. & Eichl.) Pierre, "caimitillo", with 36.48, (12.16%) of the total; *Parkia velutina* Benoist, "pashaco", with 29.33, (9.78%); *Sloanea guianensis* (Aubl.) Benth., "casha huayo", 22.04, (7.35%). These 03 species totaled 87.85, representing 29.28% of the total (Table N° 07 and Graph N° 04).

Table No. 07. Importance Value Index (IVI) of the most representative forest species of the site
IVI

absolRelat   (%)

Scientific	nameCommon name	IVI at 300%	IVI at 100% Ecclinusa
lanceolata "caimitillo".		36.48	12.16
Parkia velutina	"pashaco "	29.339	.78
Sloanea guianensis	"casha huayo "	22.047	.35
sub total87.8529		.28	

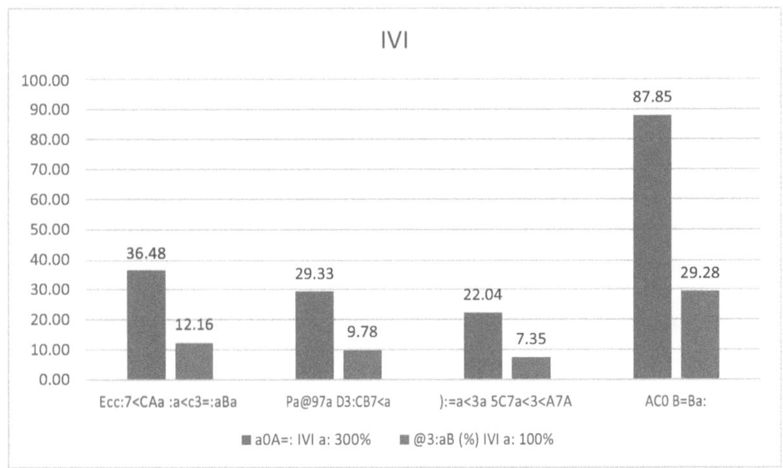

Gráfico N° 04. Índice Valor de Importancia, (IVI), de las especies forestales más representativas del lugar

Table N° 08. Importance value index (IVI), total of the most representative forest species of the site

Family	Scientific Name	Common name	IVI at 300%.	IVI at 100%
Sapotaceae	Ecclinusa lanceolata	caimitillo	36.48	12.16
Fabaceae	Parkia velutina	pashaco	29.33	9.78
Elaeocarpaceae	Sloanea guianensis	casha huayo	22.04	7.35
Araliaceae	Dendropanax umbellatus	phosphorus caspi	18.66	6.22
Urticaceae	Pourouma tomentosa	sacha uvilla	18.5	6.17
Lauraceae	Ocotea aciphylla	cinnamon moena	18.14	6.05
Euphorbiaceae	Hevea pauciflora	shiringa	17.13	5.71
Moraceae	Ficus guianensis	renaco	13.46	4.49
Lecythidaceae	Eschweilera coriacea	black machimango	11.81	3.94
Fabaceae	Inga tessmannii	shimbillo	11.81	3.94
Lauraceae	Ocotea argyrophylla	brown leaf moena	10.13	3.38
Lauraceae	Ocotea gracilis	odorless moena	9.784	3.26
Lauraceae	Ocotea myriantha	heron moena	9.41	3.14
Elaeocarpaceae	Sloanea durissima	casha huayo	8.821	2.94
Euphorbiaceae	Hyeronima oblonga	high altitude caspi mouse	8.613	2.87
Lauraceae	Ocotea oblonga	shicshi moena	8.188	2.73
Lauraceae	Anaueria brasiliensis	añuje rumo	8.157	2.72
Lauraceae	Ocotea olivacea	YELLOW SPRING	8.052	2.68
Burseraceae	Protium altsonii	copal	8.008	2.67
Fabaceae	Swartzia benthamiana	shimbillo steel	8.008	2.67
Myristicaceae	Iryanthera juruensis	red cumalilla	7.843	2.61
Fabaceae	Hymenolobium nitidum	mari mari	7.625	2.54
		Total	300	100

4.1.5 Complexity Floristic

Floristic complexity of the evaluated area was 1/1, that is, for each species there were 1 tree (Table N° 09 and Graph N° 05).

Table N° 09. Floristic complexity of the most important forest species of the site

Number of Species	1
Number of trees	1
FLORISTIC COMPLEXITY	1/1

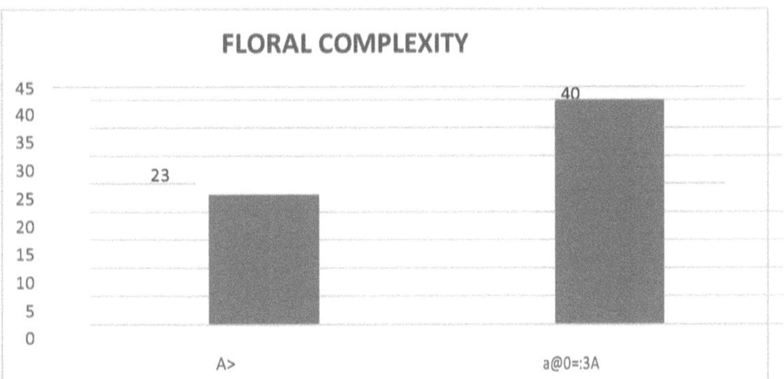

Figure N° 05. Floristic Complexity (FC) of the most representative forest species of the site.

4.1.6 Analysis of the structure horizontal (Eh)

Referred to the diameter of 40 trees, distributed in groups of diameter classes, ranging from 5 cm to 10 cm dbh, at 1.30 m from the ground.

In diameter class I, between 10 cm and 14.9 cm DBH, 13 trees (32.5%) were inventoried. Quantitatively, the following trees stood out, with 1 tree: *Sloanea guianensis* (Aubl.) Benth., "casha huayo", 14.8 cm; *Ocotea oblonga* (Meisn.) Mez, "shicshi moena", 14.6 cm; *Parkia velutina* Benoist, "pashaco", 14.5 cm. In diameter class II, between 15 cm and 19.9 cm dbh, 09 trees (22.5%) were inventoried. Quantitatively, the following trees stood out with 1 tree: *Ecclinusa lanceolata* (Mart. & Eichl.) Pierre, "caimitillo", 19.0 cm; *Hevea pauciflora* (Spruce ex Benth.) Müll. Arg., "shiringa", 19.5 cm; Hevea *pauciflora* (Spruce ex Benth.) Müll. Arg., "shiringa", 19.7 cm.

In diameter class III, between 20 cm and 24.9 cm DBH, 08 trees were inventoried (20.0%). Quantitatively, the most important, with 1 tree: *Ocotea argyrophylla* Ducke, "brown leaf moena", 24.0 m.
In diameter class IV, between 25 cm and 29.9 cm DBH, 07 trees (17.5%) were inventoried. Quantitatively larger, 1 tree, the following stood out: *Eschweilera coriacea* (A. DC.) S. A. Mori, "machimango negro", 29.80 cm; *Inga tessmannii* Harms, "shimbillo", 29.80 cm and *Ecclinusa lanceolata* (Mart. & Eichl.) Pierre,

"caimitillo", 29.28 cm.

In diameter class V, between 30 cm and 34.9 cm DBH, 02 trees were inventoried (5%), including 1 tree: *Pourouma tomentosa* Mart, "sacha uvilla", 30.5 cm and *Ficus guianensis* Desv.

In diameter class VI, between 35 cm and 39.9 cm DBH, 01 tree was inventoried (2%), including 1 tree: *Ocotea aciphylla* (Nees) Mez, "cinnamon moena", 46 cm (Table N° 10 and Graph N° 06).

Table N° 10. Horizontal structure, (Eh), of the most representative forest species of the site

DIAMETRIC CLASS	TREES (cm)	N°	(%)
I	10-14.9	13	32,5
II	15-19.9	9	22,5
III	20-24.9	8	20,0
IV	25-29.9	7	17,5
V	30-34.9	2	5,0
VI	35-39.9	1	2,5
		40	100

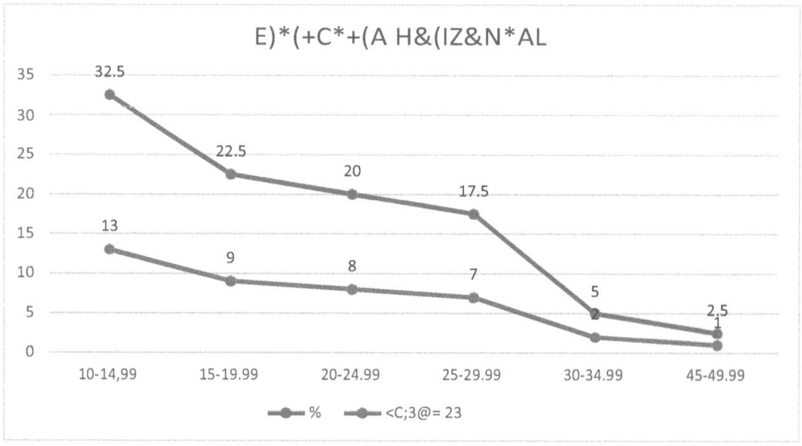

Graph N° 06. Horizontal structure of most representative forest species of the site

4.1.7 Structure analysis vertical (Ev)

Referred to the height of 40 trees, gathered in 21 different forest species, distributed in three (03) strata: lower (< 9 m), middle (10-15 m) and upper (> 15).

The lower vertical structure (EVI), less than 9 meters in total height, was found in 05 trees, (12.5%), grouped in 05 different species; *Sloanea durissima*, "casa huayo", 01 tree, 18.2 m in height; *Anaueria brasiliensis*, "añuje rumo", 01 tree, 14.4 m in height.

They presented average vertical structure (EVM), between 09 and 15 meters of total height, 22 trees, (55.0%), gathered in 14 different forest species; among others, *Ecclinusa lanceolata* (Mart. & Eichl.) Pierre, "caimitillo", 01 tree, 29.8 m high; *Parkia velutina*, "pashaco", 01 tree, 22.2 m high; *Dendropanax umbellatus*, "fosforo caspi", 01 tree, 20.4 m high.

They presented an upper vertical structure (EVS), greater than 15 meters in height, 13 trees (32.50%), comprised of 11 different forest species, including *Ocotea aciphylla*, "canela moena", 01 tree; *Ficus guianensis*, "renaco", 01 tree; *Pourouma tomentosa*, "sacha uvilla" (Table N° 11 and Graph N° 07).

Table N° 11. Vertical structure (Ev) of the most representative forest species of the site

E.V. (m)	TOTAL TREE HEIGHT RANGE	No.	(%)
BELOW	< 8	5	12.5
MEDIO	9-15	22	55.0
SUPERIOR	> 16	13	32.5
		40	100

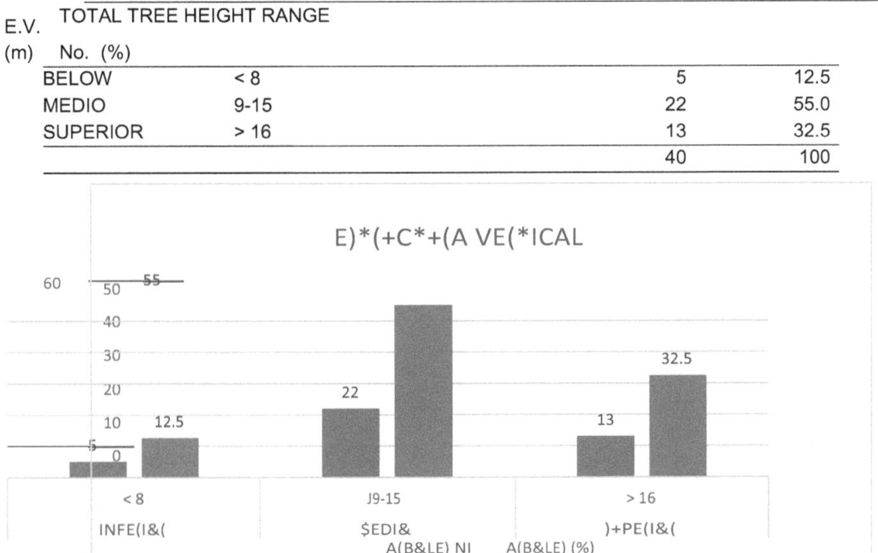

Graph No. 07. Vertical structure of the most representative forest species of the site

Table N° 12: Analysis of the horizontal and vertical structure of the most representative forest species of the site

Family	Scientific Name	Common Name	d (cm)	cd	h	ca
Lecythidaceae	Eschweilera coriacea	black machimango	29.8	25-29,99	16	15-19.99
Urticaceae	Pourouma tomentosa	sacha uvilla	27.2	25-29,99	19	15-19.99
Urticaceae	Pourouma tomentosa	sacha uvilla	30.5	30-34,99	19	15-19.99
Lauraceae	Ocotea olivacea	YELLOW SPRING	13.7	10-14,99	11	10-14.99
Fabaceae	Parkia velutina	pashaco	12.8	10-14,99	10	10-14.99
Araliaceae	Dendropanax umbellatus	caspi phosphorus	11.9	10-14,99	7	5-9.99
Elaeocarpaceae	Sloanea guianensis	casha huayo	26.0	25-29,99	15	15-19.99
Sapotaceae	Ecclinusa lanceolata	caimitillo	16.3	15-19,99	12	10-14.99
Fabaceae	Parkia velutina	pashaco	21.4	20-24,99	17	15-19.99
Araliaceae	Dendropanax umbellatus	caspi phosphorus	26.0	25-29,99	11	10-14.99
Lauraceae	Ocotea argyrophylla	brown leaf moena	24.0	20-24,99	15	15-19.99
Fabaceae	Parkia velutina	pashaco	22.2	20-24,99	14	10-14.99
Fabaceae	Swartzia benthamiana	shimbillo steel	13.4	10-14,99	6	5-9.99
Lauraceae	Ocotea aciphylla	cinnamon moena	45.5	45-49,99	22	20-24.99
Euphorbiaceae	Hevea pauciflora	shiringa	13.4	10-14,99	9	5-9.99
Araliaceae	Dendropanax umbellatus	caspi phosphorus	20.4	20-24,99	15	15-19.99
Fabaceae	Parkia velutina	pashaco	15.0	15-19,99	14	10-14.99
Sapotaceae	Ecclinusa lanceolata	caimitillo	26.3	25-29,99	17	15-19.99
Sapotaceae	Ecclinusa lanceolata	caimitillo	19.0	15-19,99	12	10-14.99
Fabaceae	Parkia velutina	pashaco	14.5	10-14,99	15	15-19.99
Elaeocarpaceae	Sloanea guianensis	casha huayo	14.8	10-14,99	11	10-14.99
Myristicaceae	Iryanthera juruensis	red cumalilla	12.2	10-14,99	13	10-14.99
Sapotaceae	Ecclinusa lanceolata	caimitillo	22.5	20-24,99	11	10-14.99
Fabaceae	Inga tessmannii	shimbillo	29.8	25-29,99	18	15-19.99
Euphorbiaceae	Hevea pauciflora	shiringa	19.7	15-19,99	14	10-14.99
Sapotaceae	Ecclinusa lanceolata	caimitillo	21.1	20-24,99	18	15-19.99
Elaeocarpaceae	Sloanea durissima	cepanchina	18.2	15-19,99	9	5-9.99
Lauraceae	Ocotea oblonga	shicshi moena	14.6	10-14,99	14	10-14.99
Fabaceae	Parkia velutina	pashaco	16.6	15-19,99	14	10-14.99
Elaeocarpaceae	Sloanea guianensis	casha huayo	18.7	15-19,99	14	10-14.99
Fabaceae	Hymenolobium nitidum	mari mari	10.4	10-14,99	13	10-14.99
Euphorbiaceae	Hevea pauciflora	shiringa	19.5	15-19,99	16	15-19.99
Lauraceae	Anaueria brasiliensis	añuje rumo	14.4	10-14,99	9	5-9.99
Lauraceae	Ocotea myriantha	heron moena	21.0	20-24,99	17	15-19.99
Moraceae	Ficus guianensis	renaco	34.6	30-34,99	23	20-24.99
Burseraceae	Protium altsonii	copal	13.4	10-14,99	14	10-14.99
Euphorbiaceae	Hyeronima oblonga	high altitude caspi mouse	17.1	15-19,99	16	15-19.99
Elaeocarpaceae	Sloanea guianensis	casha huayo	12.4	10-14,99	13	10-14.99
Lauraceae	Ocotea gracilis	odorless moena	22.6	20-24,99	16	15-19.99
Sapotaceae	Ecclinusa lanceolata	caimitillo	29.8	25-29,99	15	15-19.99

Legend: diameter (d), diameter class (cd), height (h), altimetric class (ca).

4.2. Parcel N° 02

4.2.1 Absolute abundance and relative

A total of 32 trees were inventoried, belonging to 13 families, 18 genera and 20 different species.

The most abundant local species were: *Parkia velutina* Benoist, "pasaco", 04 trees, (12.5%); *Ocotea myriantha* (Meisn.) Mez, "garza moena", 04 trees, (12.5%); *Sloanea guianensis* (Aubl.) Benth., "casa huayo", 03 trees, (9.4%), of the total. These 03 species totaled 11 trees, representing 34.4% of the total (Table N° 13 and Graph N° 08).

Table N° 13. Absolute and relative abundance of the most representative forest species of the site.

		ABUNDANCE	
Scientific Name	Common name	absol (N°)	relat (%)
Parkia velutina	"pashaco"	4	12.5
Ocotea myriantha	"heron moena"	4	12.5
Sloanea guianensis	"casha huayo"	3	9.4
sub total		11	34.4
Total		32	100

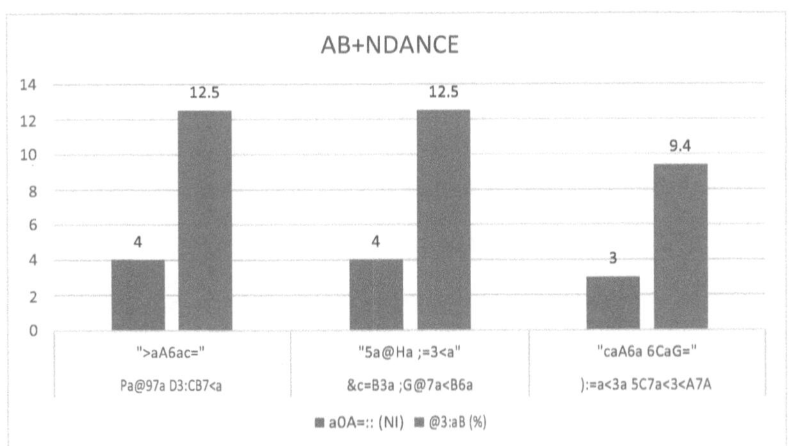

Graph N° 08. Absolute and relative abundance of the most representative forest species of the site.

4.2.2 Frequency, absolute and relative

Frequency was evaluated in 03 plots, divided into 1 subplot (0.6 ha.).

The 20 forest species each presented an absolute frequency value of 1 and a relative frequency value of 4.76% of the total (Table No. 14 and Graph No. 09).

Absolute and relative frequency of the most representative forest species of the site.

	FREQUENCY		
Scientific Name	Common Name	absol (N°)	relat (%)
Parkia velutina	"pashaco"	1	4.76
Ocotea myriantha	"heron moena"	1	4.76
Sloanea guianensis	"casha huayo"	1	4.76
		3	14.29
		32	100

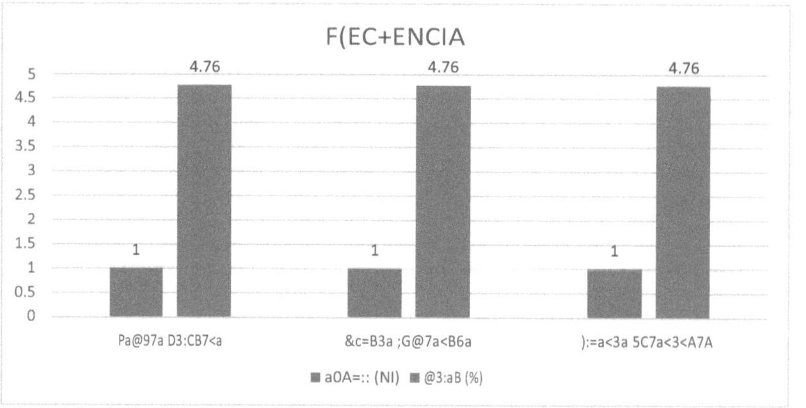

Absolute and relative frequency of the most representative forest species of the site.

4.2.3 Dominance, absolute and relative

Referred to the basal area of 32 trees, belonging to 20 different local species.

The species with the highest absolute and relative dominance were *Parkia velutina* Benoist, "pashaco", 0.2 m^2 , (27.5%); *Ocotea myriantha* (Meisn.) Mez, "garza moena", with 0.1 m^2 , (8.4%); followed by *Sloanea guianensis* (Aubl.) Benth., "casha huayo", 0.1 m^2 , (6.2%). These 03 local forest species totaled 0.4 m^2 of basal area, representing 42.1% of the total (Table N° 15 and Graph N° 10).

Table N° 15. Absolute and relative dominance (m^2) of the most representative forest species of the site.

DOMINANCE

Scientific Name	Common name	absol (m^2)	relat (%)
Parkia velutina	"pashaco"	0.2	27.5
Ocotea myriantha	"heron moena"	0.1	8.4
Sloanea guianensis	"casha huayo"	0.1	6.2
sub total		0.4	42.1
Total		0.8	100

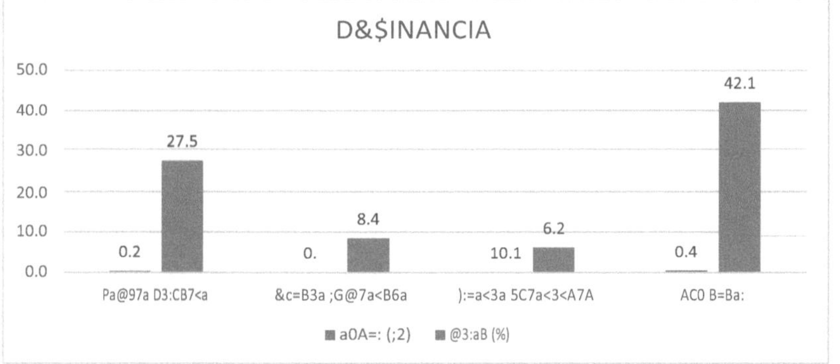

Graph N° 10. Absolute and relative species dominance (m^2), absolute and relative representative forest species of the site

4.2.4 Importance Value Index, (IVI)

A total of 20 different local species were calculated, grouping a total of 32 trees.

The species with the highest IVI were *Parkia velutina* Benoist, "pashaco", 35.4, (11.8%); *Ocotea myriantha* (Meisn.) Mez, "garza moena", 25.7, (8.6% of the total); *Sloanea guianensis* (Aubl.) Benth., "casha huayo", 23.4, (7.8%). These 03 species, totaling 84.5, represented 28.2% of the total (Table N° 16 and Graph N° 11).

Table N° 16. Importance Value Index (IVI) of the most representative forest species of the site

	IVI		
Scientific Name	Common name	IVI at 300%.	IVI at 100%
Parkia velutina	"pashaco"	35.4	11.8
Ocotea myriantha	"heron moena"	25.7	8.6
Sloanea guianensis	"casha huayo"	23.4	7.8
sub total		84.5	28.2

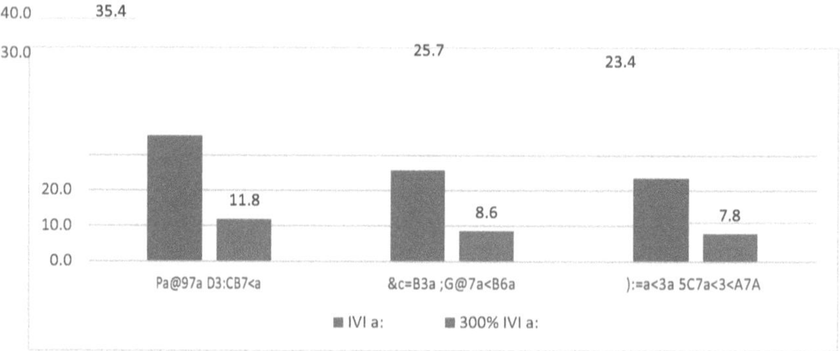

Importance Value Index (IVI), of the most representative forest species of the site.

Table N° 17. Total importance value index (IVI) of the most representative forest species of the site.

Family	Scientific Name	Common Name	IVI at 300%	IVI at 100%
Humiriaceae	Sacoglottis amazonica Mart.	shungo parrot	35.42	11.81
Lauraceae	Ocotea myriantha (Meisn.) Mez.	heron moena	25.70	8.57
Fabaceae	Parkia velutina Benoist	pashaco	23.42	7.81
Elaeocarpaceae	Sloanea guianensis (Aubl.) Benth.	casha huayo	20.31	6.77
Lauraceae	Nectandra paucinervia Coe-Teix.	moena	18.60	6.20
Annonaceae	Xylopia cuspidata Diels	caspi turtle	18.11	6.04
Sapotaceae	Ecclinusa lanceolata (Mart. & Eichl.) Pierre	caimitillo	13.67	4.56
Fabaceae	Macrolobium microcalyx Ducke	saint caspi	13.03	4.34
Lecythidaceae	Eschweilera parvifolia Mart. ex A. DC.	black machimango	12.81	4.27
Urticaceae	Pourouma mollis Trécul	sacha uvilla	12.07	4.02
Burseraceae	Protium ferrugineum (Engl.) Engl.	red copal	11.87	3.96
Fabaceae	Swartzia klugii (R.S. Cowan) Torke	sacha cumaceba	11.61	3.87
Lauraceae	Ocotea gracilis (Meisn.) Mez.	odorless moena	10.39	3.46
Meliaceae	Guarea macrophylla Vahl	requia	9.87	3.29
Lauraceae	Ocotea bracteosa (Meisn.) Mez.	cinnamon moena	9.58	3.19
Myristicaceae	Iryanthera paraensis Huber	cumalilla	9.23	3.08
Melastomataceae	Miconia symplectocaulos Pilg.	caracha caspi	8.97	2.99
Urticaceae	Cecropia latiloba Miq.	white cetico	8.89	2.96
Lauraceae	Beilschmiedia sp.	cinnamon moena	8.82	2.94
Araliaceae	Dendropanax umbellatus (Ruiz & Pav.) Decne. & Planch.	phosphorus caspi	8.82	2.94
Lecythidaceae	Eschweilera coriacea (A. DC.) S. A. Mori	black machimango	8.82	2.94
		Total	300	100

4.2.5   Floristic Complexity, (FC)

Floristic complexity of the evaluated area was 1/1, that is, for each species there was 1 tree (Table N° 18 and Graph N° 12).

Table N° 18. Floristic complexity of the most representative forest species of the site.

Number of Species	1
Number of trees	1
FLORISTIC COMPLEXITY	1/1

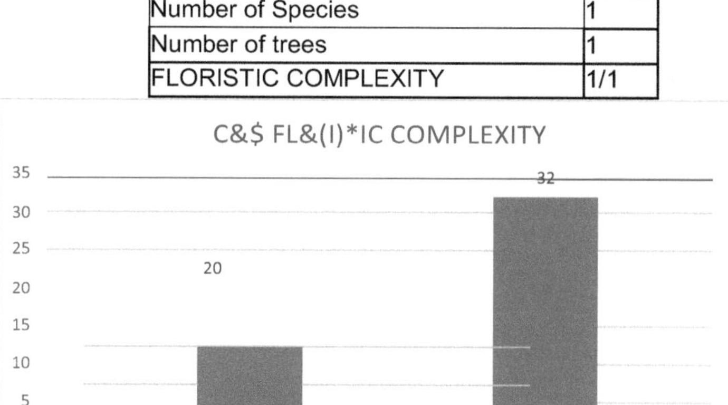

Floristic complexity of the most representative forest species of the site.

4.2.6 Analysis of the structure horizontal

Referred to the diameter of each of the 32 trees, distributed in groups of diameter classes, in ranges of 5 cm from 10 cm dbh, at 1.30 meters from the ground.

Diameter class I, between 10 cm and 14.9 cm dbh, 19 trees were inventoried (59.38%). Quantitatively larger, with 1 tree, *Guarea macrophylla* Vahl, "requia", 14.6 cm; *Ocotea myriantha* (Meisn.) Mez, "garza moena", 14.3 cm.

In diameter class II, between 15 cm and 19.9 cm DBH, 05 trees were inventoried (15.63%). Quantitatively greater, with 1 tree, were *Sloanea guianensis* (Aubl.) Benth., "casa huayo", 18.5 cm; *Ocotea myriantha* (Meisn.) Mez, "garza moena", 18.3 cm.

In diameter class III, between 20 cm and 24.9 cm dbh, 05 trees were inventoried (15.63%). Quantitatively, with 1 tree, *Macrolobium microcalyx* Ducke, "santo caspi", 23.50 cm; *Eschweilera parvifolia* Mart. ex A. DC., "machimango", 23.00 cm.
In diameter class IV, between 25 cm and 29.9 cm dbh, 02 trees were inventoried (6.25%). Quantitatively greater, with 1 tree, *Ocotea myriantha* (Meisn.) Mez, "heron moena", 26.00 cm; *Xylopia cuspidata* Diels, "espintanilla broadleaf", 25.50 cm.

In diameter class IX, between 50 cm and 54.9 cm of DBH, 01 tree was inventoried, the species *Sacoglottis amazonica* Mart, "loro shungo", 54.4 cm (Table N° 19 and Graph N° 13).

Table N° 19. Horizontal structure, (Eh), of the most representative forest species of the site.

	DIAMETER CLASS (cm)	TREES N°	(%)
I	10 - 14.9	19	59.38
II	15 - 19.9	5	15.63
III	20 - 24.9	5	15.63
IV	25 - 29.9	2	6.25
-	....	....	....
X	50 - 54.9	1	2.5

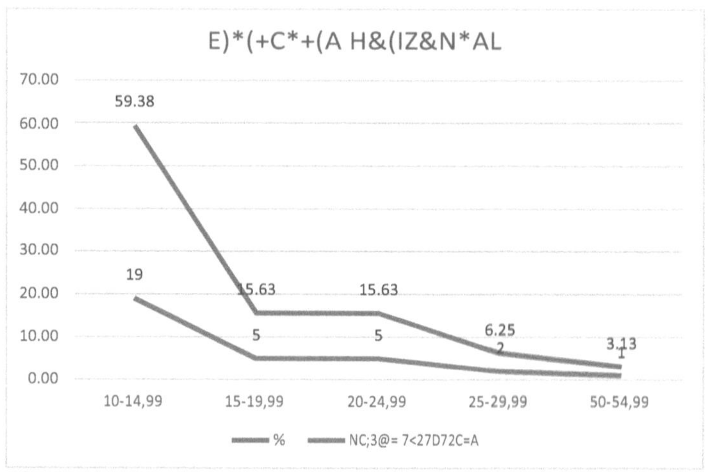

Graph N° 13. Horizontal structure of the most representative forest species at the site

4.2.7 Analysis of the structure vertical

Referred to the total height of each of the 32 trees, gathered in 20 different forest species, distributed in three (03) strata: lower (< 8), middle (9-15) and upper (> 15).

There were 07 trees (21.88%) with lower vertical structure (EVI), between 6 and 8 meters in total height, grouped in 02 different species; *Guarea macrophylla* Vahl, "requia", 8 m; *Nectandra paucinervia* Coe-Teix, "moena", 8 m; *Parkia velutina* Benoist, "pashaco", 7 m, stood out with 1 tree.
They presented medium vertical structure, (EVM), between 09 and 15 meters of total height, 21 trees (65.63%), gathered in 16 different species; among others, *Protium ferrugineum* (Engl.) Engl., "copal colorada", 15 m; *Parkia velutina* Benoist, "pashaco", 15 m; *Ocotea myriantha* (Meisn.) Mez, "garza moena", 03 trees.

There were 04 trees (12.50%) with a height of more than 16 meters, comprising 04 different species; among others, *Cecropia latiloba* Miq, "cetico blanco", 24 m; *Sacoglottis amazonica* Mart. (Table N° 20 and Graph N° 14).

Table N° 20. Vertical structure (Ev) of the most representative forest species of the site

TOTAL HEIGHT RANGE E.V. (N°) (%)		TREES (m)
BELOW	< 8	7	21.88
MEDIO	9-16	21	65.63
SUPERIOR	> 17	4	12.5

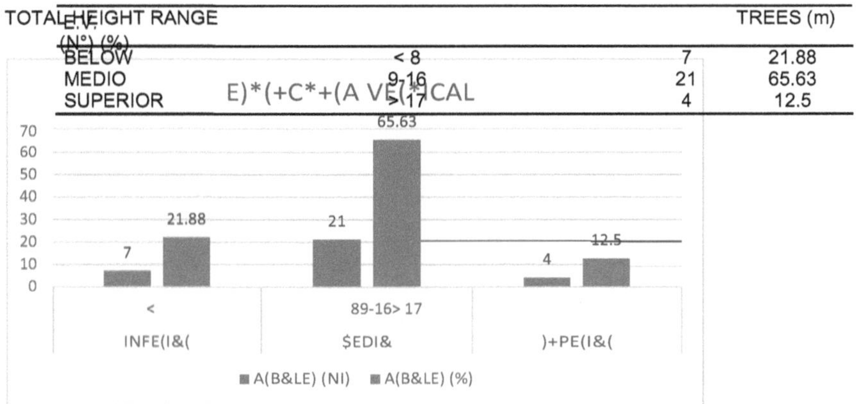

Vertical Structure, (Eh), of the most important forest species, by species. representative of the place

Table N° 21: Analysis of the horizontal and vertical structure of the most representative forest species of the site

Family	Scientific name	Name Com)n	d (cm)	cd	h	ca
LaC@ac3a3	&c=B3a 5@ac7:7A	;=3<a A7< =:=@	16.4	15-19,99	12	10-14,99
)a>=Bac3a3	Ecc:7<CAa :a<c3=:aBa	ca7;7B7::=	12.4	10-14,99	9	5-9,99
E:a3=ca@>ac3a3	):=a<3a 5C7a<3<A7A	caA6a 6CaG=	18.5	15-19,99	11	10-14,99
FaOac3a3	Pa@97a D3:CB7<a	>aA6ac=	13.2	10-14,99	15	15-19,99
BC@A3@ac3a3	P@=B7C; 43@@@C57<3C;	c=>>a: c=:=@a2=	20.7	20-24,99	15	15-19,99
FaOac3a3	Pa@97a D3:CB7<a	>aA6ac=	11	10-14,99	10	10-14,99
LaC@ac3a3	&c=B3a ; G@7a<B6a	5a@Ha ; =3<a	12.1	10-14,99	13	10-14,99
A<<=<ac3a3	XG:=>7a cCA>72aBa	B=@BC5a caA>7	10.6	10-14,99	9	5-9,99
L3cGB672ac3a3	EAc6E37:3@a c=@7ác3a	; ac67;a<5= <35@=	10	10-14,99	6	5-9,99
L3cGB672ac3a3	EAc6E37:3@a >a@D74=:7a	;ac67;a<5=	23	20-24,99	21	20-24,99
LaC@ac3a3	B37:Ac6;7327a A>.	ca<3:a ; =3<a	10	10-14,99	8	5-9,99
FaOac3a3	)Ea@BH7a 9:C577	Aac6a cC;ac30a	20	20-24,99	12	10-14,99
LaC@ac3a3	&c=B3a 0@acB3=Aa	ca<3:a ;=3<a	13.5	10-14,99	14	10-14,99
$3:7ac3a3	GCa@3a ;ac@=>6G::a	@3?C7a	14.6	10-14,99	8	5-9,99
$3:aAB=;aBac3a3	$7c=<7a AG;>:3cB=caC:=A	ca@ac6a caA>7	10.8	10-14,99	12	10-14,99
+@B7cac3a3	C3c@=>7a :aB7:=0a	c3B7c= 0:a<c=	10.4	10-14,99	24	20-24,99
A@a:7ac3a3	D3<2@=>a<aF C;03::aBCA	46A4=@= caA>7	10	10-14,99	7	5-9,99
FaOac3a3	Pa@97a D3:CB7<a	>aA6ac=	10.8	10-14,99	14	10-14,99
)a>=Bac3a3	Ecc:7<CAa :a<c3=:aBa	ca7;7B7::=	11.5	10-14,99	9	5-9,99
LaC@ac3a3	N3cBa<2@a >aCc7<3@D7a	;=3<a	26	25-29,99	16	15-19,99

39

LaC@ac3a3	N3cBa<2@a >aCc7<3@D7a	;=3<a	11.8	10-14,99	8	5-9,99
Fa0ac3a3	Pa@97a D3:CB7<a	>aA6ac=	15.8	15-19,99	7	5-9,99
Fa0ac3a3	$ac@=:=07C; ;7c@=ca:GF	Aa<B= caA>7	23.5	20-24,99	10	10-14,99
LaC@ac3a3	&c=B3a ;G@7a<B6a	5a@Ha ;=3<a	15	15-19,99	13	10-14,99
+@B7cac3a3	P=C@=C;a ;=::7A	Aac6a CD7::a	21.2	20-24,99	16	15-19,99
$G@7AB7cac3a3	I@Ga<B63@a >a@a3<A7A	cC;a:7::a	12	10-14,99	7	5-9,99
HC;7@7ac3a3	)ac=5:=BB7A a;aHó<7ca	:=@= A6C<5=	54.4	50-54,99	23	20-24,99
E:a3=ca@>ac3a3	):=a<3a 5C7a<3<A7A	caA6a 6CaG=	11.8	10-14,99	14	10-14,99
LaC@ac3a3	&c=B3a ;G@7a<B6a	5a@Ha ;=3<a	14.3	10-14,99	14	10-14,99
E:a3=ca@>ac3a3	):=a<3a 5C7a<3<A7A	caA6a 6CaG=	13.5	10-14,99	13	10-14,99
LaC@ac3a3	&c=B3a ;G@7a<B6a	5a@Ha ;=3<a	18.2	15-19,99	16	15-19,99
A<<=<ac3a3	XG:=>7a cCA>72aBa	B=@BC5a caA>7	25.5	25-29,99	17	15-19,99

Legend: diameter (d), diameter class (cd), height (h), altimetric class (ca).

4.3. Parcel N° 03

4.4.

4.4.1 Absolute abundance and relative

A total of 40 trees were inventoried, belonging to 15 families, 20 genera, 23 different local species.

The most abundant species were: *Parkia velutina* Benoist, "pashaco", 06 trees, (15.0%); *Xylopia cuspidata* Diels, "espintanilla ancha hoja ancha", 05 trees, (12.5%); *Hevea pauciflora* (Spruce ex Benth.) Müll. Arg., "shiringa", 03 trees, (7.5%). These 03 species, totaling 14 trees, represented 32.5% of the total (Table N° 22 and Graph N° 15).

Absolute and relative abundance of the most representative forest species of the site.

	ABUNDANCE		
Scientific Name	Common name	absol (N°)	relat (%)
Parkia velutina	pashaco	6	15
Xylopia cuspidata	Broadleaf shiringa	5	12.5
Hevea pauciflora		3	7.5
	sub total	14	32.5
	Total	40	100

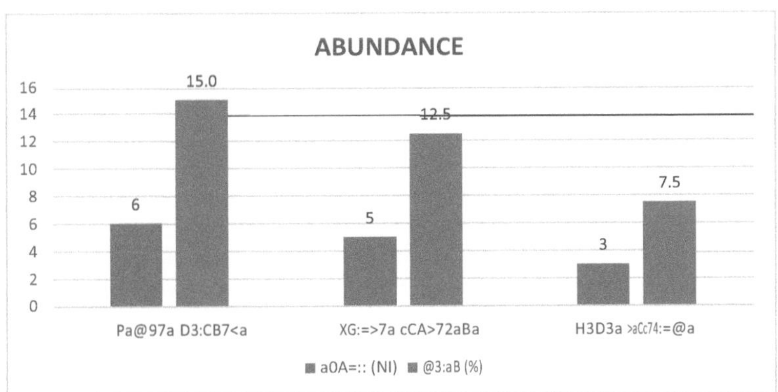

Absolute and relative abundance of the most representative forest species of the site.

4.4.2 Frequency, absolute and relative

Frequency was evaluated in 03 plots, divided into 1 subplot (0.6 ha.).

The 40 local forest species each presented an absolute frequency value of 1 and a relative frequency value of 4.17%. (Table N° 23 and Graph N° 16).

Table N° 23. Absolute and relative frequency of the most representative forest species of the site.

FREQUENCY

Scientific Name	Common Name		absol (N°)	relat (%)
Parkia velutina	pashaco		1	4.17
Xylopia cuspidata	Espintanilla broadleaf		1	4.17
Hevea pauciflora	shiringa		1	4.17
		sub total	3	12.51
		Total	24	100

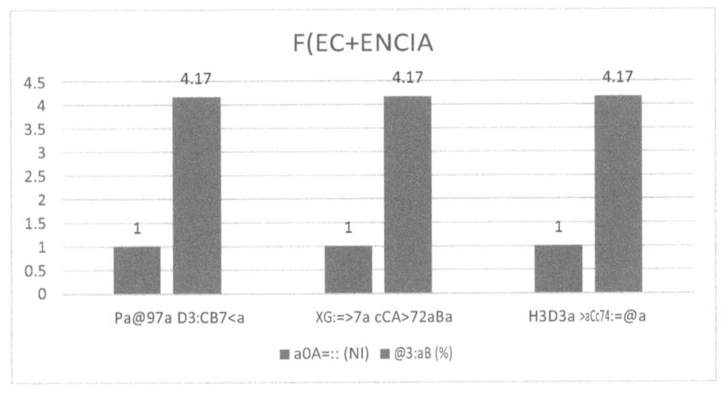

Absolute and relative frequency of the most representative forest species of the site.

4.4.3 Dominance, absolute and relative

Referred to the basal area of the 40 trees belonging to 23 different regional forest species.

The species with the highest absolute and relative dominance were *Parkia nitida* Miq., "pashaco", with 0.26 m^2 , (15.73%); followed by *Hevea pauciflora* (Spruce ex Benth.) Müll. Arg., "shiringa", 0.14 m^2 , (8.57%); *Xylopia cuspidata* Diels, "espintanilla broadleaf", 0.13 m^2 , (8.04%). These 03 local forest species totaled 0.53 m^2 of basal area, representing 32.34% of the total (Table N° 24 and Graph N° 17).

Table N° 24. Absolute and relative dominance (m^2) of the most representative forest species of the site.

Scientific name	Common name	DOMINANCE absol (N°)	relat (%)
Parkia velutina	pashaco	0.26	15.73
Hevea pauciflora	shiringa	0.14	8.57
Xylopia cuspidata	Espintanilla broadleaf	0.13	8.04
		sub total 0.53	32.34
		Total 1.67	100

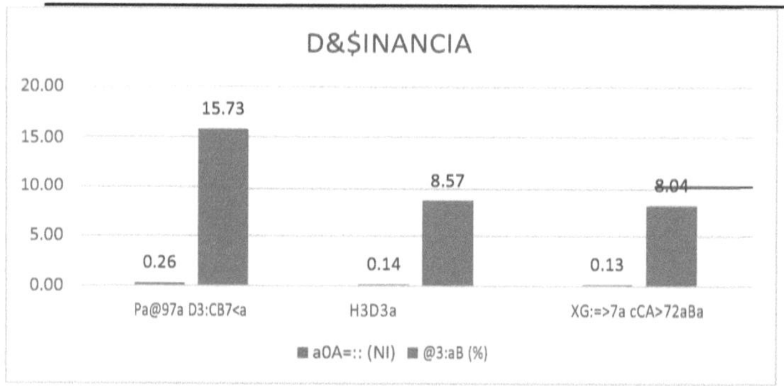

Graph N° 17. Absolute and relative species dominance (m).^{2} representative forest species of the site

4.4.4 Index Value of Importance

The Importance Value Index (IVI) of the species in the evaluated plot was calculated for 23 different species grouping a total of 40 trees. The species with the highest IVI were *Parkia velutina* Benoist, "pashaco", 35.08, (11.69%); *Xylopia cuspidata* Diels, "espintanilla broadleaf", 24.89, (8.30%); *Hevea pauciflora* (Spruce ex Benth.) Müll. Arg., "shiringa", 20.42, (6.81%). These 03 species totaled 80.39 m^2

, representing 26.80% of the total (Table N° 25 and Graph N° 18).

Table N° 25. Importance Value Index (IVI) of the most representative forest species of the site

Scientific Name	IVI Common Name		IVI at 300%.	IVI at 100%
Parkia velutina	pashaco		35.08	11.69
Xylopia cuspidata	toruga caspi shiringa		24.89	8.3
Hevea pauciflora			20.42	6.81
		sub total	80.39	26.8
		Total	300	100

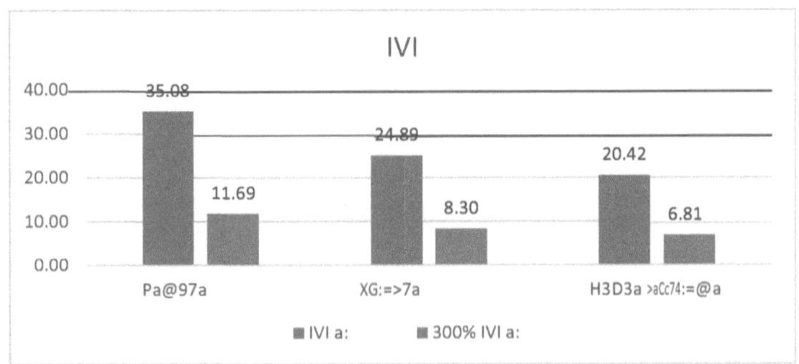

Importance Value Index (IVI), of the most representative species of the site

Table N° 26. Importance value index (IVI) of the most representative forest species of the site

Family	Scientific Name	Common Name	IVI at 300%	IVI at 100%
Fabaceae	Parkia velutina Benoist	pashaco	35.08	11.69
Annonaceae	Xylopia cuspidata Diels	caspi turtle	24.89	8.30
Euphorbiaceae	Hevea pauciflora (Spruce ex Benth.) Müll. Arg.	shiringa	20.42	6.81
Anacardiaceae	Tapirira guianensis Aubl.	wira caspi	18.12	6.04
Clusiaceae	Moronobea coccinea Aubl.	sulfur caspi	17.92	5.97
Lauraceae	Ocotea gracilis (Meisn.) Mez.	odorless moena	17.76	5.92
Fabaceae	Macrolobium microcalyx Ducke	saint caspi	17.11	5.70
Sapotaceae	Micropholis venulosa (Mart. & Eichl.) Pierre	ballet	14.70	4.90
Myristicaceae	Iryanthera tricornis Ducke	pucuna caspi	12.45	4.15
Urticaceae	Pourouma tomentosa Mart.	sacha uvilla	11.14	3.71
Apocynaceae	Macoubea guianensis Aubl.	huayo syrup	9.15	3.05
Olacaceae	Tetrastylidium peruvianum Sleumer	yutubanco	9.11	3.04
Myristicaceae	Iryanthera juruensis Warb.	red cumalilla	8.68	2.89
Myrtaceae	Calyptranthes ruiziana O. Berg	guayabilla	8.02	2.67
Lecythidaceae	Eschweilera tessmannii Knuth	machimango colorado	7.98	2.66
Sapotaceae	Micropholis egensis (A. DC.) Pierre	quinilla	7.89	2.63
Sapotaceae	Chrysophyllum bombycinum T. D. Penn.	quinilla colorada	7.78	2.59
Fabaceae	Tachigali tessmannii Harms	tangarana	7.73	2.58
Lecythidaceae	Eschweilera albiflora (A. DC.) Miers	machimango colorado	7.44	2.48
Burseraceae	Protium ferrugineum (Engl.) Engl.	red copal	7.42	2.47
Fabaceae	Inga brachyrhachis Harms	shimbillo	7.41	2.47
Malpighiaceae	Byrsonima stipulina J. F. Macbr.	sacha indano	7.40	2.47
Myrtaceae	Eugenia florida DC.	sacha guava	7.37	2.46
		Total	**300**	**100**

4.4.5 Floristic Complexity, (CF)

Floristic complexity of the evaluated area was 1/1, that is, for each species there was 1 tree (Table N° 27 and Graph N° 19).

Floristic complexity of the most representative forestry species in the site most representative species of the site

Number of Species	1
Number of trees	1
FLORISTIC COMPLEXITY	1/1

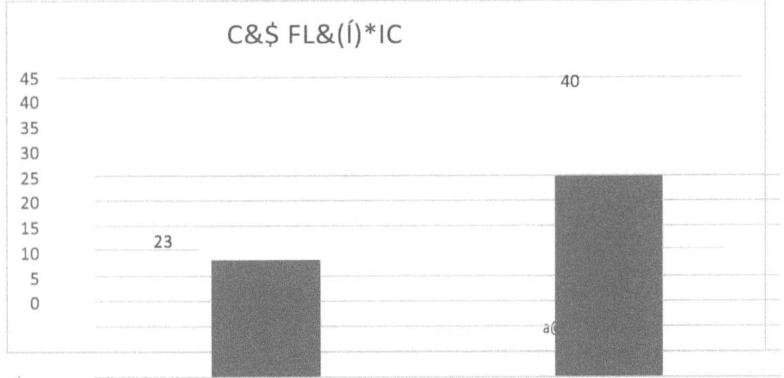

Floristic complexity of the most representative forest species of the site.

4.4.6 Analysis of the structure horizontal

The horizontal structure (Eh) of the species in the evaluated plot, referred to the diameter of 40 trees, in ranges of 5 cm from 10 cm dbh, at 1.30 meters from the ground.

Diameter class I, between 10 cm and 29.9 cm dbh, 24 trees were inventoried (60.0%). Quantitatively larger, the following trees stood out: *Xylopia cuspidata* Diels, "espintanilla broadleaf", 19.9 cm, 01 tree; *Parkia velutina* Benoist, "pashaco", 19.7 cm, 01 tree; *Iryanthera juruensis* Warb, "cumalilla colorada",
17.7 cm, 01 tree.

In diameter class II, between 20 cm and 29.9 cm dbh, 10 trees were inventoried (25.0%). Quantitatively larger, *Macrolobium microcalyx* Ducke "santo caspi", 28.4 cm, 01 tree; *Hevea pauciflora* (Spruce ex Benth.) Müll. Arg., "shiringa", 24.4 cm, 01 tree.

Diameter class III, between 30 cm and 39.9 cm dbh, 04 trees were inventoried (10.0%). Quantitatively, *Tapirira guianensis* Aubl., "wira caspi", 37.7 cm, 01 tree; *Iryanthera tricornis* Ducke, "pucuna caspi", 34.5 cm, 01 tree.
Diameter class IV, between 40 cm and 49.9 cm dbh, 01 tree was inventoried (2.5%). Quantitatively higher, *Moronobea coccinea* Aubl., "azufre caspi", 48.5 cm, 01 tree.

Diameter class V, between 50 cm and 59.9 cm DBH, 01 tree was inventoried (2.5%). *Parkia velutina* Benoist, "pashaco", 57.8 cm, 01 tree (Table N° 28 and Graph N° 20).

Table N° 28. Horizontal structure (Eh) of the most representative forest species of the site.

	DIAMETER CLASS	ÀRBOLES	
	(cm)	NO.	%
I	10-19.99	24	60
II	20-29.99	10	25
III	30-39.99	4	10
IV	40-49.99	1	2.5
V	50-59.99	1	2.5
		40	100

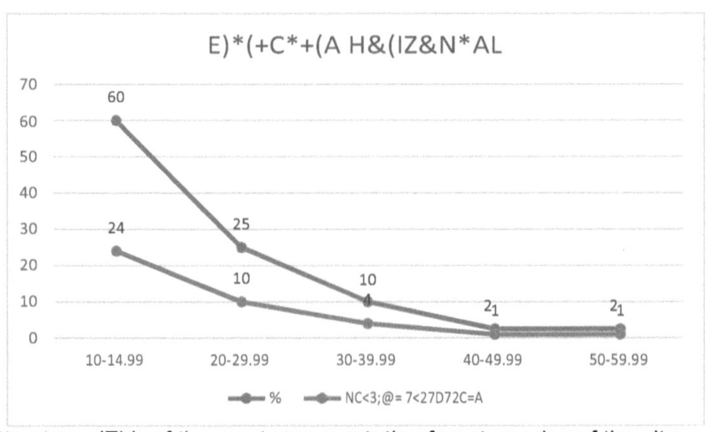

Horizontal Structure, (Eh), of the most representative forest species of the site.

4.3.7 Analysis of the vertical structure

The vertical structure (Ev) or Sociological Position of the species in the evaluated plot, referred to the total height of each of the 40 trees grouped in 23 different species.

They presented lower vertical structure (EVI), between 5 and 9.9 m of total height, 04 trees, (10.00%), gathered in 03 different species; *Protium ferrugineum* (Engl.) Engl., "copal colorado", 01 tree; *Eschweilera albiflora* (A. DC.) Miers, "machimango", 01 tree.

They presented Medium Vertical Structure (MVS), between 9 and 24 m of total height, 35 trees (87.50%), gathered in 22 different species, stood out. *Iryanthera tricornis* Ducke, "pucuna caspi", 01 tree, stood out with 24 m in height; *Moronobea coccinea* Aubl., "azufre caspi", 01 tree, with 23 m; *Macrolobium microcalyx* Ducke, "santo caspi", 01 tree, with 21 m.

They presented Superior Vertical Structure (EVS), with 25 meters high, 01 tree, *Parkia velutina* Benoist, "pashaco". (Table N° 29 and Graph N° 21).

Table N° 29. Vertical Structure, (Ev), of the most representative forest species of the site.

	TOTAL HEIGHT RANGE	TREES	
	(m)	NO.	%
I	5-9.99	7	17.5
II	10-14.99	14	35.0
III	15-19.99	15	37.5
IV	20-24.99	3	7.5
V	25-29.99	1	2.5
		40	100

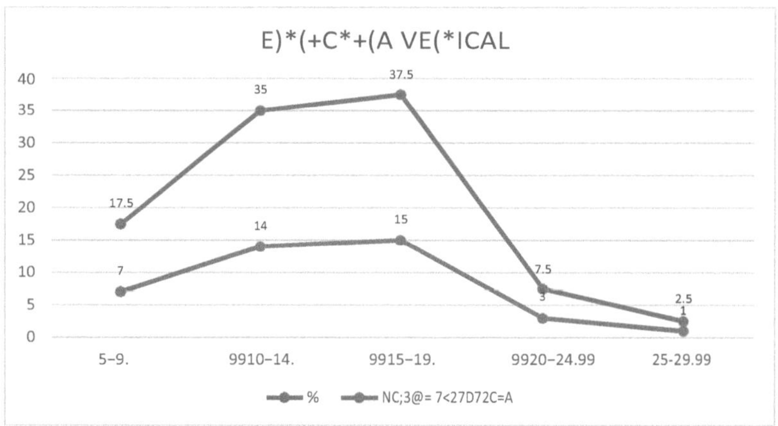

Vertical structure, (Ev), of the most important forest species (Ev). representative of the place

Table N° 30: Analysis of the horizontal and vertical structure, plot N° 03

Family	Species	Common name	d (cm)	cd	h	ca
Sapotaceae	Micropholis venulosa	ballet	17,3	10-19.99	11	10-14.99
Anacardiaceae	Tapirira guianensis	wira caspi	21,0	20-29.99	18	15-19.99
Euphorbiaceae	Hevea pauciflora	shiringa	21,7	20-29.99	16	15-19.99
Annonaceae	Xylopia cuspidata	caspi turtle	19,8	10-19.99	17	15-19.99
Annonaceae	Xylopia cuspidata	caspi turtle	19,9	10-19.99	16	15-19.99
Fabaceae	Inga brachyrhachis	shimbillo	10,9	10-19.99	14	10-14.99
Sapotaceae	Micropholis egensis	quinilla	14,9	10-19.99	12	10-14.99
Sapotaceae	Micropholis venulosa	ballet	13,6	10-19.99	13	10-14.99
Malpighiaceae	Byrsonima stipulina	sacha indano	10,8	10-19.99	9	5-9.99
Lecythidaceae	Eschweilera albiflora	machimango	11,2	10-19.99	8	5-9.99
Urticaceae	Pourouma tomentosa	sacha uvilla	30,2	30-39.99	19	15-19.99
Lecythidaceae	Eschweilera tessmannii	machimango colorado	15,5	10-19.99	11	10-14.99
Sapotaceae	Chrysophyllum bombycinum	quinilla colorada	14,1	10-19.99	12	10-14.99
Fabaceae	Parkia velutina	pashaco	15,5	10-19.99	14	10-14.99
Euphorbiaceae	Hevea pauciflora	shiringa	27,0	20-29.99	16	15-19.99
Annonaceae	Xylopia cuspidata	caspi turtle	21,0	20-29.99	15	15-19.99
Fabaceae	Parkia velutina	pashaco	14,7	10-19.99	13	10-14.99
Lauraceae	Ocotea gracilis	odorless moena	11,6	10-19.99	6	5-9.99
Fabaceae	Parkia velutina	pashaco	19,7	10-19.99	15	15-19.99
Fabaceae	Macrolobium microcalyx	saint caspi	29,0	20-29.99	17	15-19.99
Apocynaceae	Macoubea guianensis	huayo syrup	22,1	20-29.99	14	10-14.99
Sapotaceae	Micropholis venulosa	ballet	11,0	10-19.99	9	5-9.99
Lauraceae	Ocotea gracilis	odorless moena	10,8	10-19.99	7	5-9.99
Myristicaceae	Iryanthera juruensis	red cumalilla	19,7	10-19.99	14	10-14.99
Anacardiaceae	Tapirira guianensis	wira caspi	37,7	30-39.99	19	15-19.99
Olacaceae	Tetrastylidium peruvianum	yutubanco	21,9	20-29.99	16	15-19.99
Myristicaceae	Iryanthera tricornis	pucuna caspi	34,5	30-39.99	24	20-24.99
Clusiaceae	Moronobea coccinea	sulfur caspi	48,5	40-49.99	23	20-24.99
Fabaceae	Tachigali tessmannii	tangarana	13,7	10-19.99	10	10-14.99
Fabaceae	Parkia velutina	pashaco	57,8	50-59.99	25	25-29.99
Burseraceae	Protium ferrugineum	red copal	11,0	10-19.99	8	5-9.99
Lauraceae	Ocotea gracilis	odorless moena	31,7	30-39.99	15	15-19.99
Fabaceae	Macrolobium microcalyx	saint caspi	28,4	20-29.99	21	20-24.99
Myrtaceae	Calyptranthes ruiziana	guayabilla	15,8	10-19.99	13	10-14.99
Euphorbiaceae	Hevea pauciflora	shiringa	24,9	20-29.99	17	15-19.99
Fabaceae	Parkia velutina	pashaco	14,0	10-19.99	14	10-14.99
Myrtaceae	Eugenia florida	sacha guava	10,5	10-19.99	9	5-9.99
Fabaceae	Parkia velutina	pashaco	21,3	20-29.99	15	15-19.99
Annonaceae	Xylopia cuspidata	caspi turtle	12,0	10-19.99	14	10-14.99
Annonaceae	Xylopia cuspidata	caspi turtle	18,3	10-19.99	15	15-19.99

Legend: diameter (d), diameter class (cd), height (h), altimetric class (ca).

4.5. Parcel N° 04

4.5.1 Absolute abundance and relative

A total of 16 trees were inventoried, belonging to 9 families, 11 genera, 12 different species.

The most abundant local species were: *Alchornea triplinervia* (Spreng.) Müll. Arg., "zancudo caspi", 02 trees, (12.50%); *Hevea pauciflora* (Spruce ex Benth.) Müll. Arg., "shiringa", 02 trees, (12.50%); *Guarea macrophylla* Vahl, "requia", 01 tree, (6.3%). These 03 forest species, totaling 05 trees, represented 31.3% of the total (Table N° 31 and Graph N° 22).

Table N° 31. Absolute and relative abundance of the most representative forest species of the site.

Scientific nameCommon	ABUNDANCE nameabsol (N°)	relat (%)
Alchornea triplinerviazancudo	caspi212	.5
Hevea pauciflorashiringa212	.5	
Guarea macrophyllarequia16	.3	
sub total531		.3
Total16100		

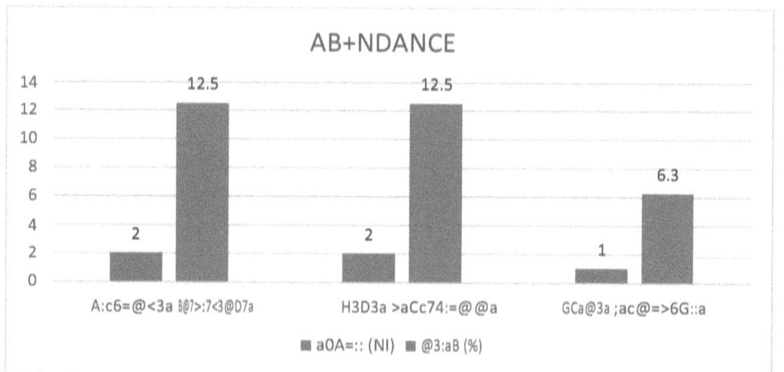

Absolute and relative abundance of the most important forest species in the country representative of the place

4.5.2 Frequency, absolute and relative

Frequency was evaluated in 03 plots, divided into 1 subplot (0.6 ha.).

The 23 local forest species each had an absolute frequency value of 1 and a relative frequency value of 8.3%. (Table N° 32 and Graph N° 23).

Table N° 32. Absolute and relative frequency of the most representative forest species of the site.

Scientific Name	Common Name	FREQUENCY Absol (N°)	Relat (%)
Guarea macrophylla	requia	1	8.3
Alchornea triplinervia	zancudo caspi	18	.3
Hevea pauciflora	shiringa	18	.3
	sub total	324	.9
	Total	16	100

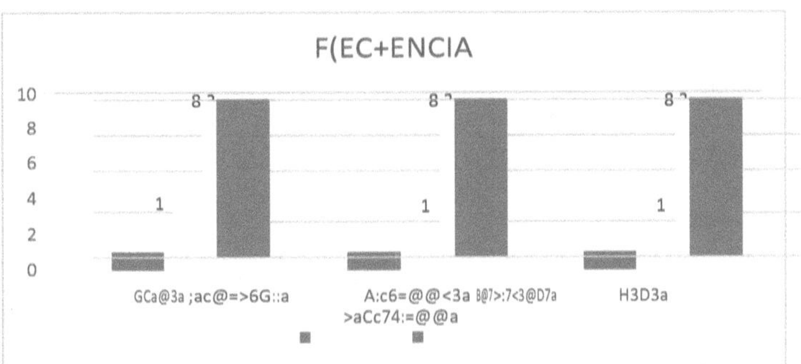

Absolute and relative frequency of the most representative forest species of the site.

4.5.3 Dominance, absolute and relative

Referred to the basal area of the 16 trees belonging to 12 different regional forest species.

Species with the highest absolute and relative dominance: *Guarea macrophylla* Vahl, "requia", 0.1 m^2 , (22.72%); *Alchornea triplinervia* (Spreng.) Müll. Arg., "zancudo caspi", 0.1 m^2 , (15.06%); *Hevea pauciflora* (Spruce ex Benth.) Müll. Arg., "shiringa", 0.1 m^2 , (11.26%). These 03 local species totaled 0.3 m^2 basal area, represented 49.04% of the total (Table No. 33 and Graph No. 24).

Table N° 33. Dominance, (m^2), absolute and relative, of the most representative forest species of the site.

Scientific Name	DOMINANCE Common Name	absol (m^2)	relat (%)
Guarea macrophylla	requia	0.1	22.72
Alchornea triplinervia	zancudo caspi shiringa	0.1	15.06
Hevea pauciflora		0.1	11.26
	sub total	0.3	49.04
	Total	16.00	100

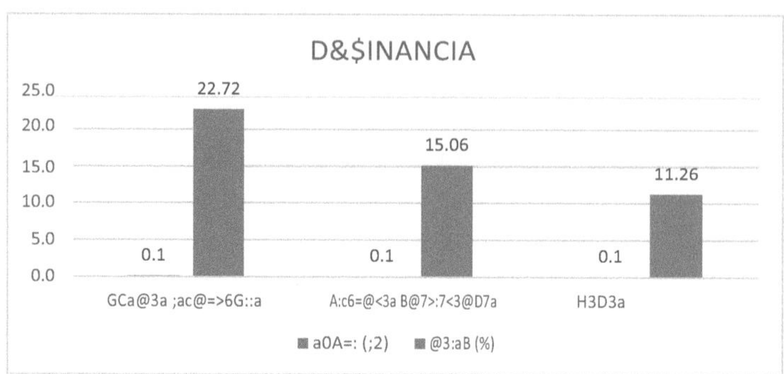

Graph N° 24. Absolute and relative species dominance (m^2), absolute and relative most representative forests of the site

4.5.4 Index Value of Importance

The Importance Value Index (IVI) of the species in the evaluated plot, 12 different species were calculated, grouping a total of 16 trees.

The species with the highest IVI were *Guarea macrophylla* Vahl, "requia", 37.31, (12.44%); *Alchornea triplinervia* (Spreng.) Müll. Arg., "zancudo caspi", 35.90, (11.97%); *Hevea pauciflora* (Spruce ex Benth.) Müll. Arg., "shiringa", 32.10, (10.70%). These 03 species totaled 105.31 IVI, representing 35.11% of the total (Table N° 34 and Graph N° 25).

Table N° 34. Importance Value Index (IVI) of the most representative forest species of the site.

Common nameSpeciesIVI at	IVI 300%IVI at 100% Guarea	macrophyllarequia
37.31		12.44
Alchornea triplinerviazancudo caspi35	.9011	.97
Hevea pauciflorashiringa32	.	1010.70
sub total105		.3135 .11
Total300100		

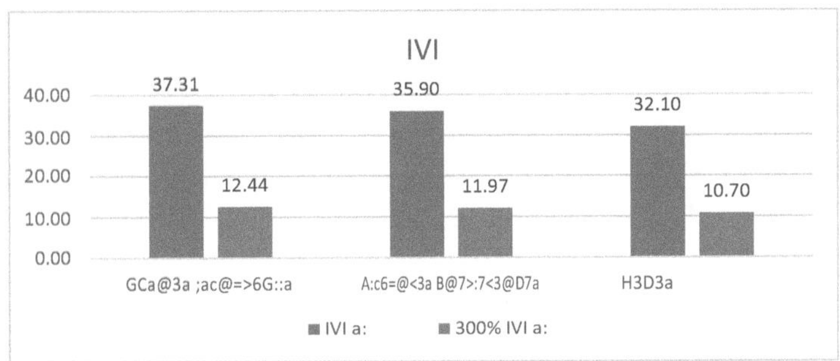

Importance Value Index, (IVI), of the most important species representative of the place

Table N° 35: Importance value index (IVI), of the most representative forest species of the site

Family	Scientific Name	Common Name	IVI at 300%	IVI at 100%
Meliaceae	Guarea macrophylla Vahl	requia	37.3	12.4
Euphorbiaceae	Alchornea triplinervia (Spreng.) Müll. Arg.	Caspian mosquito	35.9	12.0
Euphorbiaceae	Hevea pauciflora (Spruce ex Benth.) Müll. Arg.	shiringa	32.1	10.7
Myristicaceae	Osteophloeum platyspermum (A. DC.) Warb.	weeping cumala	30.5	10.2
Sapotaceae	Pouteria torta (Mart.) Radlk.	white quinilla	30.1	10.0
Annonaceae	Xylopia cuspidata Diels	caspi turtle	24.3	8.1
Apocynaceae	Aspidosperma schultesii Woodson	quillobordon	23.8	7.9
Lecythidaceae	Eschweilera coriacea (A. DC.) S. A. Mori	black machimango	19.8	6.6
Lauraceae	Aniba perutilis Hemsl.	YELLOW SPRING	17.3	5.8
Myristicaceae	Iryanthera polyneura Ducke	cumala colorada	16.6	5.5
Myristicaceae	Iryanthera juruensis Warb.	red cumalilla	16.4	5.5
Malpighiaceae	Byrsonima stipulina J. F. Macbr.	sacha indano	15.9	5.3
		Total	300	100

4.5.5 Floristic Complexity, (CF)

Floristic complexity of the evaluated area was 1/1, that is, for each species there was 1 tree (Table N° 35 and Graph N° 26).

Table   36   .   Floristic complexity of the most representative   forest species   most representative of the site

Number of Species	1
Number of trees	1
FLORISTIC COMPLEXITY	1/1

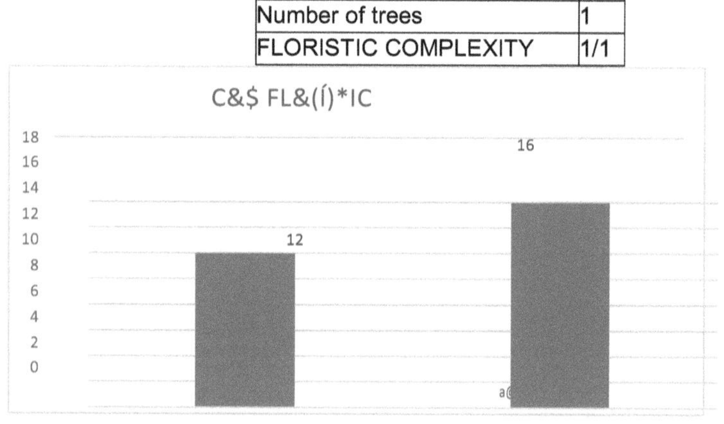

Floristic complexity of the most representative forest species of the site.

4.5.6 Analysis of the structure horizontal

The horizontal structure (Eh) of the species evaluated refers to the diameter of 16 trees, in ranges of 5 cm from 10 cm dbh, at 1.30 meters from the ground.

Diameter class I, between 10 cm and 14.9 cm DBH, 07 trees were inventoried (43.75 %). Quantitatively greater, *Aniba perutilis* Hemsl., "yellow moena", 14.3 cm, 01 tree.

In diameter class II, between 15 cm and 19.9 cm DBH, 04 trees were inventoried (25.00%). Quantitatively higher, the most outstanding were *Pouteria torta* (Mart.) Radlk., "quinilla blanca", 18.9 cm, 01 tree; *Eschweilera coriacea* (A. DC.) S. A. Mori, "machimango blanco", 19.8 cm, 01 tree.

In diameter class IV, between 25 cm and 29.9 cm DBH, 03 trees (18.75%) were inventoried. Quantitatively larger, *Alchornea triplinervia* (Spreng.) Müll. Arg., "zancudo caspi", 29.80 cm, 01 tree.

Diameter class V, between 30 cm and 34.9 cm dbh, 01 tree was inventoried; I highlight, *Osteophloeum platyspermum* (A. DC.) Warb., "weeping cumala", 34.5 cm.

Diameter class VII, between 40 cm and 49.9 cm DBH, 01 tree was evaluated, *Guarea macrophylla* Vahl, "requia", 41.2 cm (Table N° 37 and Graph N° 27).

Table N° 37. Horizontal Structure, (Eh), of the most representative forest species of the site.

DIAMETER CLASSTREES	(cm)	N O.	%
I	10-14.99	7	43.75
II	15-19.99	4	25
-	....		-
IV	25-29.99	3	18.75
V	30-34.99	1	6.25
-	....		-
VII	40-44.99	1	6.25
		16	100

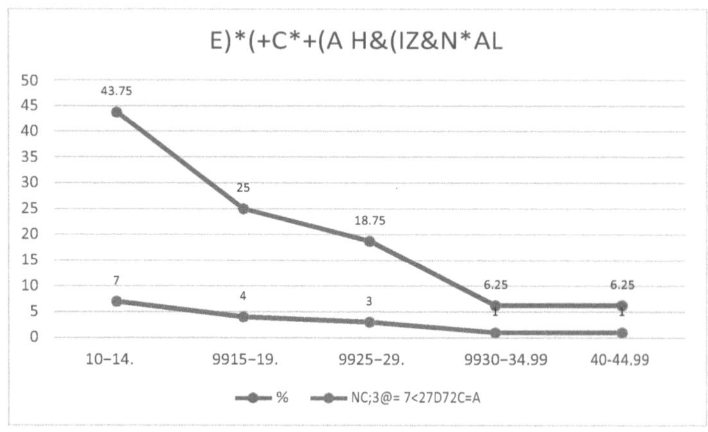

Horizontal structure (Eh) of the most representative forest species of the site.

4.5.7 Analysis of the structure vertical

The Vertical Structure (VS) or Sociological Position of the species in the evaluated plot, referred to the total height of each of the 16 trees, grouped in 12 different species.

They presented Lower Vertical Structure, (EVI), up to 6 m total height, 01 tree, (6.25%); *Iryanthera juruensis* Warb., "cumalilla colorada", stood out.

They presented Medium Vertical Structure (MVS), between 07 and 14 meters of total height, with 09 trees (56.25%), gathered in 08 different species. *Eschweilera coriacea* (A. DC.) S. A. Mori, "white machimango", 01 tree; *Pouteria torta* (Mart.) Radlk., "white quinilla", 01 tree; *Alchornea triplinervia* (Spreng.) Müll. Arg., "zancudo caspi", 01 tree.

They presented Superior Vertical Structure (EVS), more than 16 m high, 06 trees; *Osteophloeum platyspermum* (A. DC.) Warb., "weeping cumala", 01 tree, 23 meters high; *Guarea macrophylla* Vahl, "requia", 01 tree, 23 meters high; *Alchornea triplinervia* (Spreng.) Müll. Arg., "zancudo caspi", 01 tree, 22 meters high (Table N° 38 and Graph N° 28).

Table N° 38. Vertical Structure, (Ev), of the most representative forest species of the site.

TREES (m)		(No.)	(%)
E.V BELOW	< 7	1	6.25
MEDIO	8-15	9	56.25
SUPERIOR	> 16	6	37.5

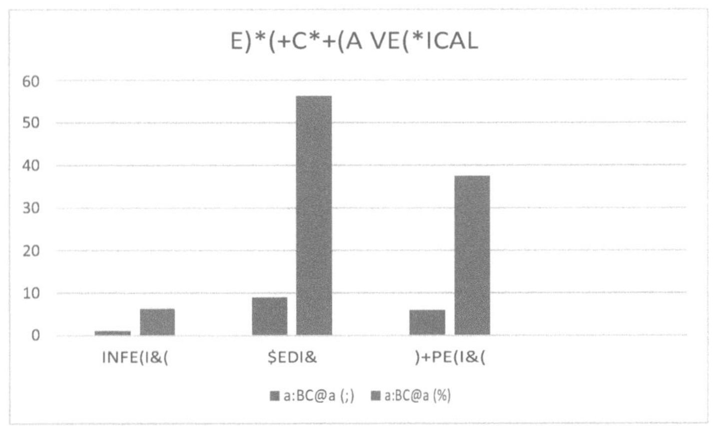

Graph N° 28. Vertical Structure, (Ev), of the most representative forest species of the site.

Table N° 39: Analysis of the horizontal and vertical structure of the most representative forest species of the site

Families	Species	Common name	d (cm)	cd	h	ca
Annonaceae	Xylopia cuspidata	broadleaf spinet	11,1	10-14.99	11	10-14.99
Annonaceae	Xylopia cuspidata	broadleaf spinet	11,5	10-14.99	13	10-14.99
Lecythidaceae	Eschweilera coriacea	black machimango	19,8	15-19.99	14	10-14.99
Sapotaceae	Pouteria torta	white quinilla	19,9	15-19.99	18	15-19.99
Meliaceae	Guarea macrophylla	requia	41,2	40-44.99	23	20-24.99
Lauraceae	Aniba perutilis	YELLOW SPRING	14,3	10-14.99	11	10-14.99
Myristicaceae	Iryanthera polyneura	cumala colorada	12,4	10-14.99	13	10-14.99
Malpighiaceae	Byrsonima stipulina.	sacha indano	10,0	10-14.99	9	5-9.99
Myristicaceae	Iryanthera juruensis	red cumalilla	11,5	10-14.99	6	5-9.99
Euphorbiaceae	Hevea pauciflora	shiringa	27,0	25-29.99	16	15-19.99
Euphorbiaceae	Hevea pauciflora	shiringa	10,6	10-14.99	8	5-9.99
Apocynaceae	Aspidosperma schultesii	quillobordon	26,2	25-29.99	16	15-19.99
Sapotaceae	Pouteria torta	white quinilla	17,2	15-19.99	14	10-14.99
Euphorbiaceae	Alchornea triplinervia	Caspian mosquito	29,8	25-29.99	22	20-24.99
Myristicaceae	Osteophloeum platyspermum	weeping cumala	34,5	30-34.99	23	20-24.99
Euphorbiaceae	Alchornea triplinervia	Caspian mosquito	15,4	15-19.99	14	10-14.99

Legend: diameter (d), diameter class (cd), height (h), altimetric class (ca).

4.6. Parcel N° 05

4.6.1 Absolute abundance and relative

A total of 24 trees were inventoried, belonging to 11 families, 15 genera, 16 different forest species.

The most abundant local species: *Alchornea triplinervia* (Spreng.) Müll. Arg., "zancudo caspi", 04 trees, (16.7%); *Swartzia benthamiana* Miq., "sacha cumaceba", 03 trees, (12.5%); *Nealchornea yapurensis* Huber, "huira caspi", 01 tree, (4.2%). These 03 species totaled 08 trees, which represented 33.4% of the total (Table N° 40 and Graph N° 29).

Table N° 40. Absolute and relative abundance of the most representative forest species of the site.

	ABUNDANCE		
Common name	Species	absol (N°)	relat (%)
Alchornea triplinervia	Caspian mosquito	4	16.7
Swartzia benthamiana	sacha cumaceba	3	12.5
caspi	Nealchornea yapurensis	1	4.2
	sub total	8	33.4
	Total	24	100

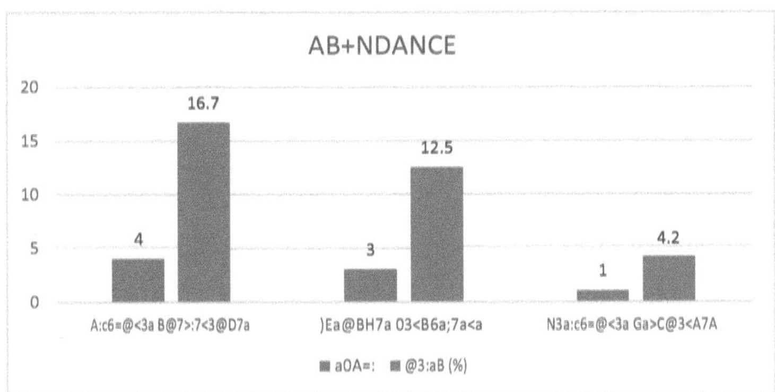

Graph N° 29. Absolute and relative abundance of the most important forest species. representative of the place

4.6.2 Absolute frequency and relative frequency

Frequency was evaluated in 03 plots, divided into 1 subplot (0.6 ha.).

The 16 local forest species each presented an absolute frequency value of 1 and a relative frequency value of 6.25% (Table No. 41 and Graph No. 30).

Table N° 41. Absolute and relative frequency of the most representative forest species of the site.

	FREQUENCY	
Common nameSpeciesabsolrelat (%)		
Alchornea triplinerviazancudo	caspi16	.25
Swartzia benthamianasacha	cumaceba16	.25
Nealchornea	yapurensishuira	caspi16 .25
sub total318		.75
Total24100		

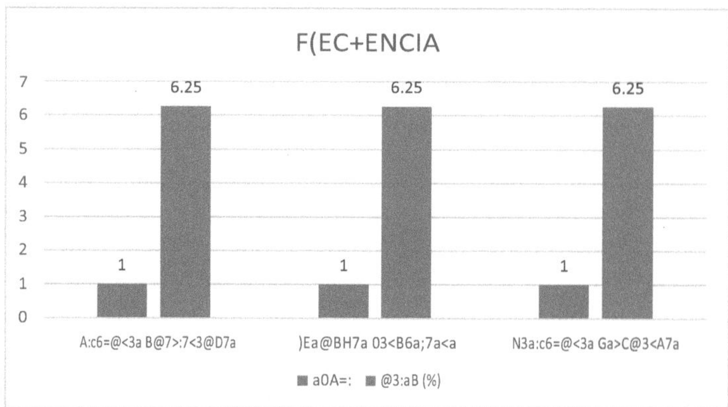

Graph N° 30. Absolute and relative frequency of the most representative forest species of the site.

4.6.3 Absolute dominance and relative

Absolute and relative dominance of forest species in the evaluated area, referring to the basal area of 24 trees, belonging to 15 different local species.

The most dominant species: *Swartzia benthamiana* Miq., "sacha cumaceba", 0.37 m^2 , (35.56%); *Alchornea triplinervia* (Spreng.) Müll. Arg.; "zancudo caspi", 0.21 m^2 , (19.81%); *Nealchornea yapurensis* Huber; "huira caspi", 0.16 m^2 , (15.26%). These 03 local species totaled 0.74 m^2 basal area, representing 70.63% of the total (Table N° 42 and Graph N° 31).

Table N° 42. Absolute and relative dominance (m^2) of the most representative forest species of the site.

Common name	Species	absol (m^2)	DOMINANCE relat (%)	
	Swartzia benthamiana	sacha cumaceba	0.37	35.56
Alchornea	triplinervia	zancudo caspi	0.2119	.81
Nealchornea	yapurensis	huira caspi	0.1615	.26
sub total		0.7470	.63	
Total		1.05100		

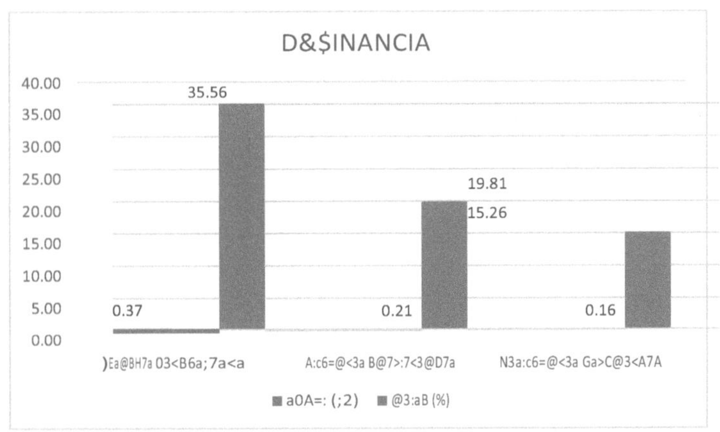

Graph N° 31. Absolute and relative dominance (m^2) of the most representative forest species of the site.

4.6.4 Importance Value Index, (IVI)

The Importance Value Index (IVI) of the species in the evaluated plot, 15 different local species were calculated, grouping a total of 24 trees.

The species with the highest IVI were, *Swartzia benthamiana* Miq., "sacha cumaceba", 54.31, (18.10%); *Alchornea triplinervia* (Spreng.) Müll. Arg., "zancudo caspi", 42.73, (14.24%); *Nealchornea yapurensis* Huber, "huira caspi", 25.67, (8.56%). These 03 species totaled 122.71 IVI, which represented 40.90% of the total (Table No. 43 and Graph No. 32).

Table N° 43. Importance Value Index (IVI) of the most representative forest species of the site.

Common nameSpeciesIVI at	IVI 300%IVI at 100% Swartzia	benthamianasacha	
cumaceba	54.31	18.10	
Alchornea triplinerviazancudo caspi42	.7314	.24	
Nealchornea	yapurensishuira caspi25	,678	.56
sub total122		.7140	.90
Total300100			

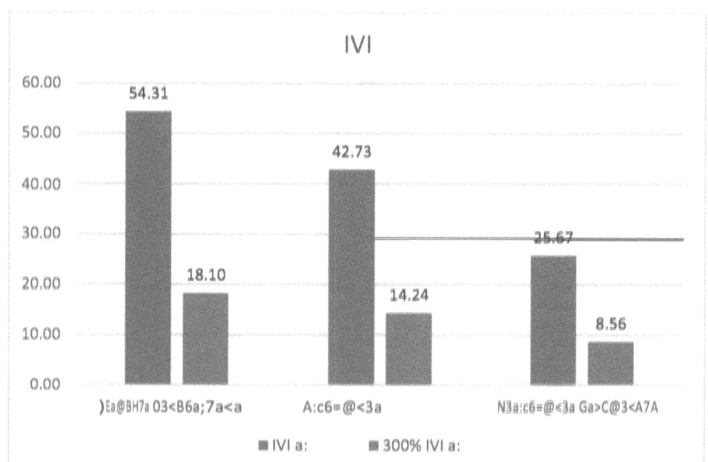

Figure N° 32. Importance Value Index, (IVI), of the most important species in the representative of the place

Table N° 44: Importance value index (IVI), of the most representative forest species of the site

Family	Scientific Name	Common Name	IVI at 300%	IVI at 100%
Fabaceae	Swartzia benthamiana Miq.	shimbillo steel	54.31	18.10
Euphorbiaceae	Alchornea triplinervia (Spreng.) Müll. Arg.	Caspian mesquite	42.73	14.24
Euphorbiaceae	Nealchornea yapurensis Huber	caspi	25.67	8.56
Icacinaceae	Dendrobangia boliviana Rusby	avocado caspi	23.01	7.67

Sapotaceae	Ecclinusa lanceolata (Mart. & Eichl.) Pierre	caimitillo	21.24	7.08
Urticaceae	Pourouma minor Benoist	sacha uvilla	15.98	5.33
Fabaceae	Parkia velutina Benoist	pashaco	13.17	4.39
Myristicaceae	Iryanthera lancifolia Ducke	cumala colorada	12.21	4.07
Myristicaceae	Iryanthera juruensis Warb.	red cumalilla	11.88	3.96
Burseraceae	Protium ferrugineum (Engl.) Engl.	red copal	11.62	3.87
Sapotaceae	Pouteria cladantha Sandwith	quinilla	11.49	3.83
Lauraceae	Ocotea gracilis (Meisn.) Mez.	odorless moena	11.47	3.82
Sapotaceae	Micropholis guyanensis (A. DC.) Pierre	ballast	11.44	3.81
Burseraceae	Crepidospermum prancei Daly	white copal	11.32	3.77
Annonaceae	Xylopia cuspidata Diels	espintanilla broadleaf	11.24	3.75
Arecaceae	Socratea exorrhiza (Mart.) H. Wendl.	casha pona	11.21	3.74
		Total	**300**	**100**

4.6.5 Complexity Floristic

Floristic complexity of the evaluated area was 1/1, that is, for each species there was 1 tree (Table No. 45 and Graph No. 33).

Table N° 45. Floristic complexity of the most representative forest species of the site

Number of Species	1
Number of trees	1
FLORISTIC COMPLEXITY	1/1

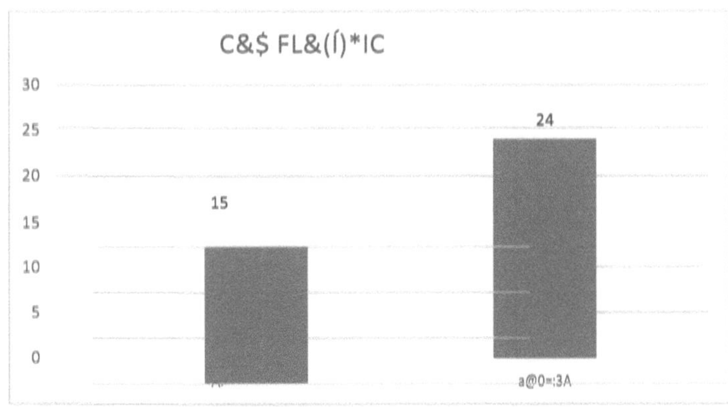

Figure N° 33. Floristic Complexity of the most representative forest species of the site.

4.6.6 Analysis of the structure horizontal

The horizontal structure (Eh) of the species in the evaluated plots is referred to the diameter of 24 trees, in ranges of 10 cm from 10 cm dbh, at 1.30 meters from the ground.

Diameter class I, between 10 cm and 19.9 cm dbh, 16 trees were inventoried (66.67%). Quantitatively larger, *Parkia velutina* Benoist, "pashaco", 19.2 cm, 01 tree; *Ecclinusa lanceolata* (Mart. & Eichl.) Pierre, "caimitillo", 16.0 cm, 01 tree; *Iryanthera lancifolia* Ducke, "cumala colorada",
15.5 cm, 01 tree.

In diameter class II, between 20 cm and 29.9 cm DBH, 04 trees (16.67%) were inventoried. Quantitatively larger, *Alchornea triplinervia* (Spreng.) Müll. Arg., "zancudo caspi", 27.3 cm, 01 tree; *Ecclinusa lanceolata* (Mart. & Eichl.) Pierre, "caimitillo", 25.2 cm, 01 tree.

Diameter class III, between 30 cm and 39.9 cm dbh, 01 tree was inventoried (4.16%). Quantitatively larger, *Alchornea triplinervia* (Spreng.) Müll. Arg., "zancudo caspi", 39.4 cm, 01 tree.

Diameter class IV, between 40 cm and 49.9 cm dbh, 02 trees were inventoried (8.33%); *Pourouma minor* Benoist, "sacha uvilla", 27.3 cm, 01 tree; *Alchornea triplinervia* (Spreng.) Müll. Arg., "zancudo caspi", 27.3 cm, 01 tree; *Ecclinusa lanceolata* (Mart. & Eichl.) Pierre, "caimitillo", 25.2 cm, 01 tree.

Diameter class V, between 50 cm and 59.9 cm DBH, 01 tree was inventoried (4.35%); *Swartzia benthamiana* Miq. "sacha cumaceba", 51.0 cm, 01 tree (Table N° 46 and Graph N° 34).

Table N° 46. Horizontal Structure, (Eh), of the most representative forest species of the site.

	DIAMETER CLASS (cm)	TREES N°	TREES (%)
I	10-19.99	16	66.67
II	20-29.99	4	16.67
III	30-39.99	1	4.16
IV	40-49.99	2	8.33
V	50-59.99	1	4.17
		24	100

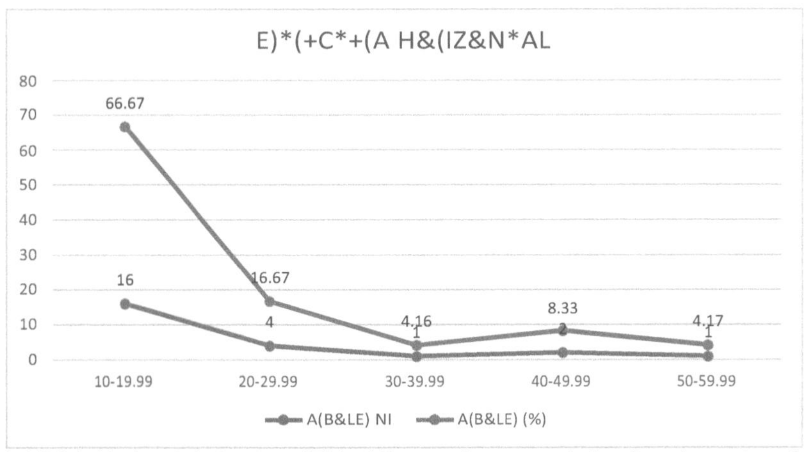

Graph N° 34. Horizontal structure (Eh) of the most representative forest species of the site.

4.5.7 Analysis of the vertical structure

The vertical structure (Ev) or sociological position of the species in the evaluated plot is referred to the total height of each of the 24 trees grouped in 15 different species.

They presented lower vertical structure, (EVI), up to 9 meters of total height, 06 trees, (25.0%); *Iryanthera juruensis* Warb., "cumalilla colorada", 22 meters high, 01 tree, stood out.

They presented medium vertical structure, (EVM), between 09 and 14 meters of total height, 10 trees stood out (41.7%), gathered in 8 different forest species; *Ecclinusa lanceolata* (Mart. & Eichl.) Pierre, "caimitillo", 14 meters high, 01 tree; *Iryanthera lancifolia* Ducke, "cumala colorada", 14 meters high, 01 tree.

They presented superior vertical structure (EVS), with more than 15 m height, 08 trees, (33.3%); *Swartzia benthamiana* Miq., "sacha cumaceba", 24 m height, 01 tree; *Nealchornea yapurensis* Huber, "huira caspi", 23 m height, 01 tree; *Alchornea triplinervia* (Spreng.) Müll. Arg., "zancudo caspi", 23 m high, 01 tree (Table N° 47 and Graph N° 35).

Tabla N° 47. Estructura Vertical, (Ev), de especies forestales más representativas del lugar

	TOTAL HEIGHT RANGE		TREES	
	m	N°		%
BELOW	< 9	6		25.0
MEDIO	9-14	10		41.7
SUPERIOR	> 15	8		33.3
		24		100

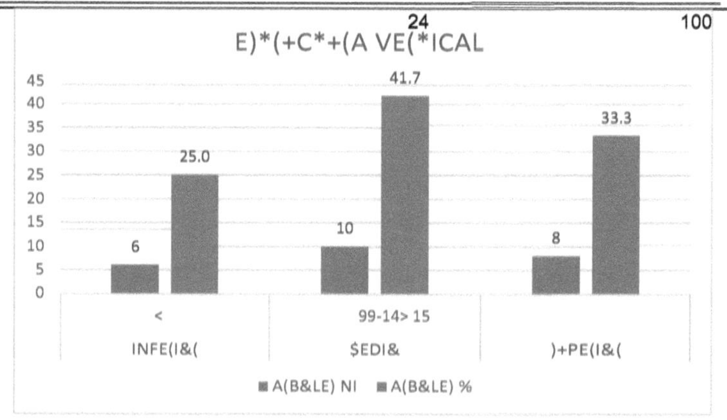

Graph N° 35. Vertical structure (Ev) of the most representative forest species of the site.

Table N° 48: Analysis of the horizontal and vertical structure of the most representative forest species of the site

Family	Scientific Name	Name Com)n	d (cm)	cd	h	ca
EC>6=@07ac3a3	A:c6=@<3a B@7>:7<3@D7a	Ha<cC2= caA>7	27.3	20-29.99	17	15-19.99
EC>6=@07ac3a3	A:c6=@<3a B@7>:7<3@D7a	Ha<cC2= caA>7	15	10-19.99	13	10-14.99
EC>6=@07ac3a3	A:c6=@<3a B@7>:7<3@D7a	Ha<cC2= caA>7	11.4	10-19.99	14	10-14.99
LaC@ac3a3	&c=B3a 5@ac7:7A	;=3<a A7< =::@	11.9	10-19.99	9	5-9.99
FaOac3a3	)Ea@BH7a 03<B6a;7a<a	Aac6a cC;ac30a	40.4	40-49.99	22	20-24.99
$G@7AB7cac3a3	I@Ga<B63@a :a<c74=:7a	cC;a:a c=:=@a2a	15.5	10-19.99	14	10-14.99
+@B7cac3a3	P=C@=C;a ;7<:@	Aac6a CD7::a	27.3	20-29.99	20	20-24.99
EC>6=@07ac3a3	A:c6=@<3a B@7>:7<3@D7a	Ha<cC2= caA>7	39.4	30-39.99	23	20-24.99
A@3cac3a3	)=c@aB3a 3F=@@@67Ha	caA6a >=<a	10.3	10-19.99	7	5-9.99
FaOac3a3	)Ea@BH7a 03<B6a;7a<a	ac3@= A67;07::=	51	50-59.99	24	20-24.99
FaOac3a3	)Ea@BH7a 03<B6a;7a<a	ac3@= A67;07::=	23	20-29.99	12	10-14.99
A<<=<ac3a3	XG:=>7a cCA>72aBa	B=@BC5a caA>7	10.5	10-19.99	12	10-14.99
Icac7<ac3a3	D3<2@=0a<57a 0=:7D7a<a	C;a@7c7::=	11.5	10-19.99	7	5-9.99
BC@A3@ac3a3	P@=B7C; 43@@@C57<3C;	c=>>a: c=:=@a2=	12.7	10-19.99	5	5-9.99
BC@A3@ac3a3	C@3>72=A>3@;C; >@a<c37	c=>a: 0:a<c=	11	10-19.99	9	5-9.99
Icac7<ac3a3	D3<2@=0a<57a 0=:7D7a<a	C;a@7c7::=	15.4	10-19.99	14	10-14.99
Icac7<ac3a3	D3<2@=0a<57a 0=:7D7a<a	C;a@7c7::=	14.2	10-19.99	13	10-14.99
)a>=Bac3a3	$7c@=>6=:7A 5CGa<3<A7A	0a:aBa @=Aa2a	11.7	10-19.99	12	10-14.99
)a>=Bac3a3	P=CB3@7a c:a2a<B6a	qC7<7::a	12	10-19.99	11	10-14.99
FaOac3a3	Pa@97a D3:CB7<a	>aA6ac=	19.2	10-19.99	19	15-19.99
EC>6=@07ac3a3	N3a:c6=@<3a Ga>C@3<A7A	6C7@a caA>7	45.2	40-49.99	23	20-24.99
$G@7AB7cac3a3	I@Ga<B63@a 8C@C3<A7A	cC;a:7::a c=:=@a2a	14	10-19.99	9	5-9.99
)a>=Bac3a3	Ecc:7<CAa :a<c3=:aBa	ca7;7B7::=	16	10-19.99	14	10-14.99
)a>=Bac3a3	Ecc:7<CAa :a<c3=:aBa	ca7;7B7::=	25.2	20-29.99	16	15-19.99

Legend: diameter (d), diameter class (cd), height (h), altimetric class (ca).

Illustration N° 06:   Abundance   profile of the most important forest species in the evaluated area.

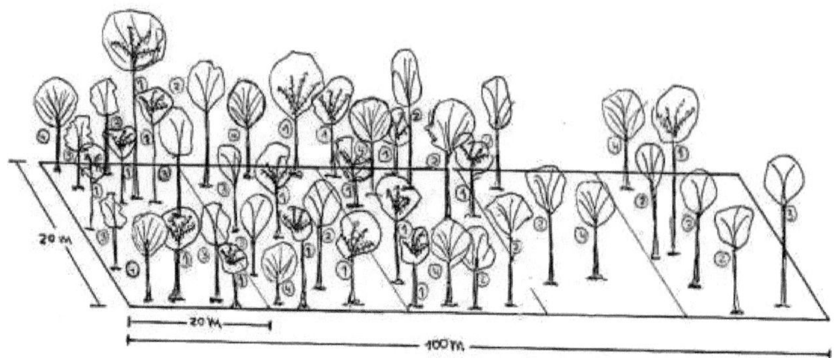

Prepared by: Dario Davila Paredes, (2024). CAPTION:
FABACEAE: *Parkia velutina*, "pashaco", 17 trees (1). SAPOTACERAE: *Ecclinusa lanceolata*, "caimitillo", 10 trees (2). ANNONACEAE: *Xylopia cuspidata*, "tortuga caspi", 10 trees (3). EUPHORBIACEAE: *Hevea pauciflora*, "shiringa", 8 trees (4).

Illustration N° 07: Profile of vertical structure of the most important forest species in the evaluated area.

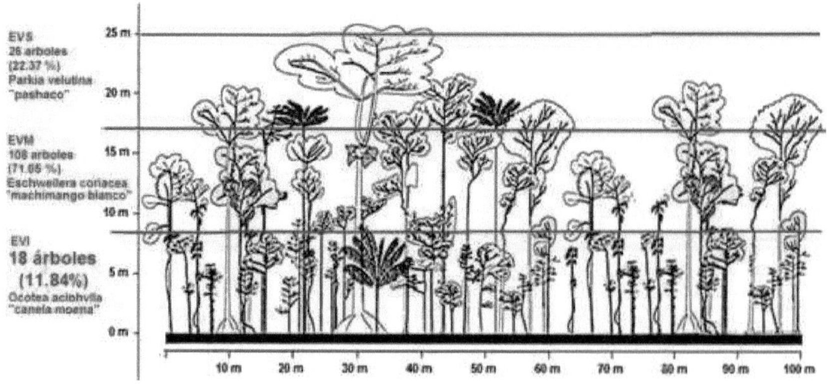

V. DISCUSSION OF THE RESULTS

1.   The applied method satisfactorily fulfills the requirements to be considered in other forest inventories, 6 000 m plots^2 (0.6 ha.), in Amazonian forests, which can be understood through small local inventories[98, 55, 58] , related to abundance, frequency, dominance (IVI), floristic complexity and horizontal and vertical structure.

2.   The area (0.6 ha), recorded 152 trees, represented 23 families, 43 genera, 64 forest species; important and representative families of the mid-terrace forest, with the largest number of genera, species: Fabaceae, 6 genera, 8 species, 29 trees, "pashaco"; Lauraceae, 5 genera, 11 species, 21 trees, "moena"; Sapotaceae, 4 genera, 7 species, 19 trees, "caimitillo"; Euphorbiaceae, 4 genera, 4 species, 16 trees, "shiringa"; Annonaceae, 1 genus, 1 species, 10 trees, "espintanilla broadleaf". These 5 families totaled 95 trees (62.5%) of the total; confirmed results, in the floristic text of three reserves of Iquitos, the 10 botanical families present in Amazonian forest inventories: Fabaceae, Lauraceae, Annonaceae, Rubiaceae, Moraceae, Myristicaceae, Sapotaceae, Meliaceae, Arecaceae and Euphorbiaceae, contribute (74.6%) and, contribute 73% of species in samples of 0.1 ha.; however, 8 of 10 families are always present in the Amazon forest and contribute the greatest number of species[55, 58, 57] ; the Fabaceae, Euphorbiaceae, and Myristicaceae families are the most representative in the Allpahuayo-Mishana National Reserve[102, 99] ; these results were also verified with exsiccata that exist in the AMAZ, collections of different authors.

3.   In three plots (0.6 ha.), the most abundant family was Fabaceae, represented by *Parkia velutina*, 17 trees, (11.18%), coinciding with the inventory of horizontal and vertical structure of two types of forest conducted in the Madre de Dios Region, among others, the most abundant species was Fabaceae, *parkia velutina* "pashaco", 17 trees, representing 70.43% of the total,[44] ; another investigation referred to the structure and floristic composition of plant communities, on the Iquitos-Nauta highway, in 5 plots (1 ha.), 1,502 different trees of 10 cm DBH were recorded, with Fabaceae being the most important family, *Inga dumosa* "shimbillo", representing 16.85% of the total number of individuals evaluated .[43]

4.   Regarding frequency, in 4 plots, the most important family and species was Fabaceae, *Parkia velutina*, "pashaco", with absolute frequency 4, (40%), coincides with what was recorded in a research work: structural analysis of a medium terrace forest in the Allpahuayo-Mishana National Reserve, the family with the highest frequency was Fabaceae, *Inga dumosa* .[103]

5.   The 3 species with the highest dominance were *Parkia velutina* "pashaco", 0.59 m^2 , *Swartzia benthaminana*, "acero shimbillo", 0.39 m^2 ; *Ecclinusa lanceolata*, "caimitillo", 0.33, m^2 and *Tapirira guianensis*, "huira caspi", 0.31 m^2 ; coincides with what was recorded in the study on structural analysis of a mid-terrace forest in the

allpahuayo-mishana national reserve, with the greatest dominance was Fabaceae, *inga dumosa*,[103] ; another inventory on horizontal and vertical structure in two types of forest, Madre de Dios Region, the most dominant species was Fabaceae family, *parkia velutina* "pashaco", with 6.61 m^2 , of 1502 trees .[44]

6.   The Importance Value Index (IVI) of 64 different species were: Fabaceae, genera *Parkia, Macrolobium* and *Swartzia*, "pashaco", "sacha cumaceba", Sapotaceae, genus *Ecclinusa*, "caimitillo", Annonaceae, genus *Xylopia,* "espintanilla broadleaf"; Lauraceae, genus *Ocotea*, "moena"; Euphorbiaceae, genera *Hevea* and *Alchornea*, "shiringa" and "zancudo caspi"; Urticaceae, genus *Pourouma*, "sacha uvilla". 08 species, represented 43.3% of the total. This result coincides with the IVI, found in the research work on structural analysis of a mid-terrace forest, Allpahuayo Mishana National Reserve, which Fabaceae, *Inga sp.*, recorded 11 species and represented 50% of the total[43] . In addition, in a research study on ecological and economic zoning, upper Amazon, it was determined that the most important species was *Parkia igneiflora*, "goma pashaco", represented by 16.45% .[89, 101]

7.   Floristic complexity was 1/3, that is, for each species there were 3 trees. Coincidentally, in Amazonian forests, the floristic mixture or complexity coefficient varies between 1:3 and 1:4 .[24]

8.   To characterize the horizontal and vertical structure of the trees, the total height of all trees was considered (e.g. Killeen *et al.*, 1988 and Lamprecht, 1990). In horizontal structure, the highest percentage of 152 trees was recorded in diameter class I (10-14.9 cm DBH), with 65 trees, representing 42.76%; a result that coincides with the structural analysis of a medium terrace forest, Allpahuayo-Mishana National Reserve, where the highest percentage (58.06%) was found distributed in class II (10-19.9 cm DBH),[43] . Regarding the vertical structure, 3 strata were established: lower, middle and upper. The middle class (9-17m), registered 108 trees (71.05%), gathered in 8 different species; *Eschweilera coriacea*, "machimango blanco"; *Pouteria torta*, "quinilla blanca"; *Alchornea triplinervia*, "zancudo caspi" stood out; a result that coincides with a study on the structural analysis of a middle terrace forest, carried out in the Allpahuayo Mishana National Reserve, the middle vertical structure (10-14.9 m), *Inga* sp. and *Eschweilera* sp. are the most abundant species, representing 2.32% of the total, [100]

VI. CONCLUSIONS

1.   In 1 hectare (10 000 m^2), 3 plots were designed, 20 m x 100 m, divided into 5 plots of 20 m x 20 m and subdivided into plots of 10 m x 10 m, 5 m x 5 m and 2 m x 2 m, (0.6 ha.).

2.   The floristic diversity of 152 trees was grouped into 23 families, 43 genera and

64 different species.

3.   Abundance: *Parkia velutina*, 17 trees (10.53%), *Ecclinusa lanceolata*, "caimitillo", 10 trees, (6.58%); *Xylopia cuspidata*, "espintanilla ancha hoja ancha", 10 trees, (6.58%); *Hevea pauciflora*, "siringa", 8 trees, (5.26%); *Sloanea guianensis*, "casa huayo", 7 trees, (4.61%). These 05 main forest species totaled 52 trees, representing 34.21% of the total.

4.   Species frequency, in 04 plots, recorded 4 different species (40%), absolute frequency 4, highlighted *Parkia velutina*; in 03 plots, recorded 05 different species (30%), absolute frequency 3, highlighted *Ecclinusa lanceolata*; in 02 plots, 14 different species (20%), absolute frequency 2, highlighted *Ocotea aciphylla*, in 01 plot, 33 local species (10%), absolute frequency 1, highlighted *Anaueria brasiliensis*. In total there were 56 different species.

5.   Dominance, referring to the basal area of 152 trees; it was Fabaceae, genus *Parkia*, totaling a subtotal of 1.61 m² basal area, represented 28.72% of the total.

6.   Species with the highest IVI: *Parkia velutina*, "pashaco", (8.7%); among others, it represented 43.3% of the total.

7.   Floristic complexity was 1/3, that is, for each species there were 3 trees.

8.   Horizontal structure, referred to diameter; diameter class I, (10-14.9 cm DBH), registered 65 trees, (42.76 %). Quantitatively greater, *Micropholis egensis*, "quinilla" 14.9 cm, 01 tree; diameter class X, (55-59.9 cm DBH), quantitatively greater *Parkia velutina*, "pashaco", 57.8 cm, 01 tree, of the total.

9.   Vertical structure, referred to total height, of 152 trees, gathered in 64 different species, the lower stratum (09-17m), quantitatively greater, with 108 trees, (71.05%); was *Eschweilera coriácea*, (14 m high), "machimango blanco".

VII.   RECOMMENDATIONS

1.   Before harvesting and reforesting the forest, first evaluate the IVI, to determine which forest species are ecologically adapted to the natural conditions of the area and allow optimal growth of the species in the lowland forest of the Peruvian Amazon.

2.   To carry out similar research work along the different Amazon basins, to obtain basic and detailed data, for the successful sustainable management of this natural resource, for the development of the Loreto region.

3.   Raise awareness of the strengths and weaknesses of the different types of vegetation ecosystems, so that their use is sustainable.

VIII. REFERENCES BIBLIOGRAPHIC REFERENCES

1.   MINAM. Memoria Descriptiva del Mapa de Ecozonas del Inventario Nacional Forestal y de Fauna Silvestre del SERFOR. National Forestry and Wildlife Service. 2019. 14 p.

2.   MEJÍA CARHUANCA, K. Diagnosis of plant resources of the Peruvian Amazon. Technical Document N° 16. Peruvian Amazon Research Institute. Iquitos - Peru. October. 1995. 4-9 p.

3.   National Forest and Wildlife Service (2019). Report of the National Forest and Wildlife Inventory of Peru. National Forestry and Wildlife Service. General Directorate of Forestry and Wildlife Information and Management. Inventory and Valuation Directorate. December. 5, p.

4.   ALVIS GORDO, JOSE F. (2009). Structural Analysis of a natural forest area located in the rural municipality of Popayán. Faculty of Agricultural Sciences. TULL Research Group. University of Cauca.

5.   BRAKO, L. & J. ZARUCCHI. Catalogue of the Flowering Plants and Gymnosperms in Peru. Monographs in Systematic Botany from the Missouri Botanical Garden, 45. St. Louis. 1993. 1286 p.

6.   BENAVIDES, M. Peruvian Amazon. Protected Areas and Indigenous Territories [notes for or map]. Instituto del Bien Común. 2009. 1 p.

7.   Vásquez Martínez, R. and Rojas Gonzáles, R. del P. Plants of the Peruvian Amazon. Key to Identify the Families of Gymnospermae and Angiospermae. ARNALDOA. Journal of the Museum of Natural History. Private University Antenor Orrego. 2006. 23 p.

8.   MACIEL-MATA, C.A., N. MARQUEZ-MORÁN, P., OCTAVIO-AGUILAR & G. SANCHEZ ROJAS. The range of species distribution: Review of the concept. Acta Univeristaria. N° 25(2). 2015. 15 p.

9.   KREBS, J. 1989. Ecology Methodology. Haroer & Row, Publishers, New York.

10. ARANA, V. F. Study of the floristic structure of forest species in a sector of the forest of plot 105 AACD "El Paujil", Sector of the Allpahuayo-Mishana National Reserve". Iquiros-Peru. 2016. 25 p.

11. ROJAS, M. V. W.; ESTEVES-VARON, J.V.; RONCANCIO, N. Structure and floristic composition of tropical rainforest remnants in eastern Caldes. Colombia. Scientific Bulletin of Natural History. Vol. 12, 2008. 24-37 pp.

12. RUIZ TAMANI, WILLER E. Floristic Composition and Horizontal Structure of the Arboreal Vegetation of the Arboretum "El huayo", Loreto-Peru. Thesis for the degree of Forestry Engineer. School of Professional Formation of Forestry Engineering. Faculty of Forestry Sciences. 2018. v. p.

13. ERAZO ARÉVALO, POOL S. Forest Composition, Horizontal Structure and Diversity of Slightly Dissected Low Hill Forests, in the Area of the Iquitos - Nauta Highway, Loreto, Peru. 2014. Thesis for the degree of Engineer in Tropical Forest Ecology. School of Professional Training of Engineering in Tropical Forest Ecology. Faculty of Forestry Sciences. UNAP. 2019. v. p.

14. Alvarez-Montalván, C. E., Manrique-León, S., Vela-Da Fonseca, M., Cardoza-Soarez, J., CAllo-Ccorcca, J., Bravo-Camara, P., Castañeda Tinco, I., & Alvarez-Orellana, J. Floristic composition, structure and tree diversity of an Amazonian forest in Peru. *Scientia Agropecuaria*. 2021.12 (1), 73-82 p.

15. ALARCÓN MOZOMBITE, EDWARD J. Composition, Structure and Floristic Diversity in Nine Localities of the Loreto Region, 2011 (Northeastern Peru). Thesis for the professional degree of Biologist. School of Professional Formation of Biology. Faculty of Biological Sciences. UNAP. 2021. xi. p.

16. SABOGAL, C. Ecological-silvicultural characterization study of the "copal" forest in Jenaro Herrera (Loreto-Peru). Thesis for the degree of Forestry Engineer. National Agrarian University La Malina. Lima. 1980. 379 p.

17. FINOL, H. Forestry in the Venezuelan Orinoquia. Mérida, Venezuela, Universidad de los Andes. Faculty of Forestry Sciences. 1974. 29 p.

18. LAMPRECHT, H. Structure and function of South American Forests. From: Ecosystem research in South America, Biogeographica, v.B. The Hague.1977.

19. SABOGAL, C. Ecological-silvicultural characterization study of the "copal" forest in Jenaro Herrera (Loreto-Peru). Thesis for the degree of Forestry Engineer. National Agrarian University La Molina. Lima. 1980. 379 p.

20. GENTRY, A. Floristic and phytogeographic diversity of Amazonia. Proceedings of the International Symposium on Amazonian Research and Management. Book 1. INDERENA. Colombia, 1989. 65 - 70 pp.

21. RICHARDS, P. The Tropical rain forest, an ecological study. Cambridge University. 1966. 450 p.

22. HOLDRIDGE, L. Ecology based on life zones. Inter-American Institute of Agricultural Sciences (IICA). San José, Costa Rica. Trad. H. Jimenez. 1978. 216 p.

23.   LOUMAN, B. Ecological bases. In: Louman Bastiaan, David Quiros Dávila, and Margarita Nilson (editors). Silviculture of broadleaf forests with emphasis on Central America. Turrialba - Costa Rica. Serie técnica. Technical Manual/CATIE: No. 46. 2001. 265 pp.

24.   LAMPRECHT. Silviculture in the tropics; forest ecosystems in tropical forests and their tree species - possibilities and methods for sustainable utilization. Institute of Forestry, University of Gottingen - Germany. Translated by Antonia Garrido, Germany. 1990. 335 pp.

25.   GENTRY, A. Tree species richness of the upper Amazonian forest. Proc. Natl. Acad. Sci. 1988. 85: 156-159 p.

26.   GENTRY, A. Summary of neotropical phytogeographic patterns and their implications for the development of Amazonia. Revista Academia Colombiana de Ciencias 1986. 85: 101-116 p.

27.   ESCOBAR, N. Diagnosis of the Floristic Composition Associated to Agricultural Activities in Cerro Quinini (Colombia). Agricultural Sciences Journal. University of Cundinamarca. 2013. 1 p.

28.   Villarreal, H., Álvarez, M., Córdoba, S., Escobar, F., Fagua, G. and Gast, F. Manual de métodos para el desarrollo de inventarios de biodiversidad. Biodiversity Inventory Program. Grupo de Exploración y Monitoreo Ambiental (GEMA). Alexander von Humboldt Biological Resources Research Institute. Bogotá, Colombia. 2004.

29.   AGUIRRE, Z. Guide for studies of floristic composition, structure and diversity of natural vegetation. San Francisco Xavier University of Chuquisaca, Sucre, Bolivia. 2010.

30.   FONSECA, K. Restoration of vegetation cover in the Monte Alto-Hojancha Forest Reserve. Costa Rican Institute. Turrialba-Costa Rica. 2002. 95 p.
31.   LAMPRECHT, H. Essay on the floristic structure of the southeastern part of the university forest "El caimital" Rev. *Forestal venezolana*. N° 7(10). 1990. 77-119 p.

32.   QUESADA, M. Floristic and structural composition of a primary forest. School of Forest Engineering. Technological Institute of Costa Rica. 2000. 15 p.

33, BUDOWSKI, G. Conservation as an instrument for development. [Anthology]. San José, Costa Rica. Editorial UNED. 1985. 398 p.

34.   FINEGAN, B. and SABOGAL. C. The development of sustainable production systems in humid tropical lowland forests: a case study of Costa Rica. El Chasqui

(CR), 1988. N°17. 1988. 3-24 p.

35. HUTCHINSON, I. D. Points of departure for silviculture in humid tropical forests. Common wealth Forestry Review (G.B.). N° 67(3). 1988. 223-230 p.

36. MAGURRAN, A. E. Ecologycal diversity and its measuremet. Princeton University Press, New Jersey. 1988.

37. KREBS, C. J. *Ecological methodology.* Harper Collins Publ. 1989. 654 pp.

38. LAMPRECHT, H. Essay on the floristic structure of the southeastern part of the university forest "El Caimital". Barinas State. Venezuelan Forestry Magazine. N° 6. 1964. 10-11 p.

39. Sabogal, C.; Carrera, F.; Colan, V.; Pokorny, B. & Louman B. Manual for the Planning and Evaluation of Operational Forest Management in Peruvian Amazon Forests. INRENA-CIFOR-FONDEBOSQUE Project. Lima, Peru. 2004.

40. SABOGAL, C. Structure and Dynamics of a Forest in the Pucallpa Region (Peruvian Amazon). Institute of Tropical Silviculture and Natural Forest Research. UNALM. Lima-Peru. 1983. 31 p.

41. AMARAL, P.; VERISSIMO, A.; BARRETO, P. & VIDAL, E. Bosque para Siempre. Manual for Timber Production in the Amazon. WWF. USAID. SIDA. Talleres Gráficos de Artegrafía SRL. Lima-Peru. 2005. 161 p.

42. MINAM. National Map of Vegetation Cover. Memoria descriptiva/ Ministerio del Ambiente, Dirección General de Evaluación, Valoración y Financiamiento del Patrimonio Natural. Lima: MINAM, 2015.

43. ZÁRATE, R.; T. MORI; J. MACO. Structure and floristic composition of plant communities in the area of the Iquitos-Nauta Highway, Loreto, Peru. Folia Amazónica N° 22(1-2). 2013. 77-89 p.

44. QUISPE VILLAFUERTE, W. "Horizontal and vertical structure of two types of concession forest in the Madre de Dios Region". Thesis for the degree of forestry engineer. National Amazonian University of Madre de Dios Environment. Puerto Maldonado - Peru. 2010.

45. Forestry Promotion Department. Forest management: Development of management plans and operational plans for harvesting in tropical rainforests. National Forestry Institute. Nicaragua. 2006.

46. MORENO LOZANO, J. M. Horizontal Structure and Economic Valuation of commercial timber species in four types of forests, Torres Causana District

(Undergraduate Thesis). National University of the Peruvian Amazon - Peru. 2015.

47. OROZCO, L.; C, BRUMER. 2002. Forest inventory for broadleaved forests in Central America. Technical Series, (CATIE) N°50. Turrialba (Costa Rica), 2002, pp. 35-68.

48. KREBS, CH. J. Ecology of Distribution and Abundance. Trans. by Jorge Blanco Correa. 2 ed. Mexico. Industria Editorial. 1985.

49. FRANCO, L.J., FIGUEROA, E., CARRASCO, A. Y TORRES, J. Manual de Ecología 2 reimp. Mexico Editorial trillas, S.A. de C.V. 1989.

50. UNESCO. World Conference on Cultural Policies (MONDIACULT). 1982.

51. AGUILAR FLORES, J.A. Horizontal structure and timber volume in forests along the Iquitos - Nauta highway, Loreto, Peru". Thesis to obtain the degree of forestry engineer professional training school of forestry engineering. Faculty of Forestry Sciences. Iquitos-Peru. 2014. 30 p.

52. UGALDE, L. Basic Concepts of Dasometry. Swiss Development Cooperation Program (DDA), Forestry Information and Documentation for Tropical America, Turrialba -Costa Rica. 1981.

53. PARDÉ, J.; BOUCHON, J. Dasometria. 2ed. Trans. from the French by Antonio Prieto Rodríguez. Madrid, Spain. Editorial: Paraninfo S. A. 1994.

54. GÓMEZ, D. Floristic composition in the riparian forest of the upper San Alberto watershed. Oxapampa. Thesis. Universidad Nacional Agraria La Molina, Lima-Peru. 2000. 59 p.

55. GENTRY, A. Summary of neotropical phytogeographic patterns and their implications for the development of Amazonia. Revista Academia Colombiana de Ciencias 1986. 85: 101-116 p.

56. VÁSQUEZ MARTÍNEZ, R. Flora of the Biological Reserves of Iquitos, Peru. Allpahuayo-Mishana, Explornapo Camp, Explorama Lodge. Edited by Agustín Rudas Lleras and Charlotte M. Taylor. Sponsored by the John D. and Catherine T. MacArthur Foundation. Additional support from Instituto de Investigaciones de la Amazonía Peruana (IIAP). Explorama Tours S. A. Missouri Botanical Garden. 1997.

57. AYALA FLORES, F. Plant Taxonomy. Gymnospermae and Angiospermae of the Peruvian Amazon. Printed by CETA. Vol. I and II, Edition I. 2003.

58. VASQUEZ, R. Flora of the Biological Reserves of Iquitos - Peru. Monographs in Systematic Botany from the Missouri Botanical Garden. Vol. No. 63. Missouri

Botanical Garden Press. St. Louis 1046 pages. 1997.

59. CHEESSON, P & T. CASE. Overview. Nonequilibrium Community theories: change, variability, history, and coexistence. In Community Ecology. Harper and Row. N. York. p 229- 238. 1986.

60. FOSTER, R. & S. HUBBEL. Vegetation structure and species composition of a fifty hectare plot on Barro Colorado Island in Tropical Forest Ecology. Seasonal cycles and long-term change. Pg. 129-140. 1990.

61. THORINGTON, R. B. TANNENBAUM, A. TARAK & R. RUDRAN. Tree distribution on Colorado mud island: a five hectare sample. In Ecology of a tropical forest. Seasonal cycles and long-term change. Egbert G, Stanley R. & D. Windson editors. Smithsonian Tropical Research. Institute, balboa, Panama. Pg. 129-140. 1990.

62. Valencia, R. H. Basley & G. Paz y Minoc. High tree alpha-diversity in amazonian Ecuador. Biodiversity and Conservation 3: 21-28. 1994.

63. Spichiger, R. *et al.* Tree species richness of a South-West Amazonian Forest (Jenáro Herrera, Perú, 73.40' W'/4.54' S). Candollea 51 (2): 559-577. 1996.
64. https://reefresilience.org.

65. MALLEUX, J. 1982. Forest Inventory in Tropical Forests. Lima- Peru, 193 p.

66. LAMPRECH, H. 1962. Tests for methods of structural analysis of tropical forests. Acta Científica Venezolana. 13(2): 57-65.

67. PACHECO, T. and TORRES, J. Dispersion Analysis of Twelve Forest Species of CIEFOR - Puerto Almendra. National University of the Peruvian Amazon. Iquitos-Peru. 51 p.

68. MINISTRY OF AGRICULTURE. Explanatory Guide to the Forest Map. INRENA. Lima-Peru. 25 p.

69. ONERN. 1976. Ecological map of Peru. Descriptive guide. Lima-Peru. 1- 146 p.

70. BRAKKO, L. & J. ZARUCCHI. 1993. Catalogue of the Flowering Plants and Gymnosperms in Peru. Monographs in Systematic Botany from the Missouri Botanical Garden, 45 St. Louis. 1286 p.

71. VASQUEZ MARTINES, R. 1997. Florula of the Biological Reserves of Iquitos, Peru. Allpahyuayo-Mishana, Explornapo Camp, Explorama Lodge. Missouri Botanical Garden.

72.   MATTEUCCI, S. & COLMA, A 1982. Methodology for the study of vegetation. Venezuela, p. 99.

73.   FRANCO, J. *et. al.* 1995. Manual de ecología Editorial Trillas. Third Reprint. 266 p.

74.   FREITAS, E. 1986. "Influence of timber harvesting on the structure and floristic composition of a high riparian forest in Jenaro Herrera-Peru". Thesis for the degree of Forestry Engineer-UNAP/FIF. Iquitos. 171 p.

75.   ROLLET, B. 1994. L' architecture das Forest denses humides sempervirentes de plaine Centre Technique Forestier Tropical, Nogent. Sur Mana. France. 298 p.

76.   MARMILLOD, D. 1982. Methodik und ergeb nisse Von Untersuchungen Uber Zusammensetzumg und Autban eines Terrassenwaldes in peruanischen amazonien. Dissertation. Goftingen, Germany, Georg-Aujust-Universrtat Gottingen. 198 p.

77.   TELLO, C. 1995. Ecological characterization by the method of the sextants of the arboreal vegetation of a forest type varíllal of the zone of Puerto Almendras, Iquitos-Peru. Thesis for the degree of Forestry Engineer. MJNAP/FIF. Iquitos, Peru. 104 p.

78.   ZUÑIGA, G. 1985. Structural analysis of an intervened forest in the Arto Shori-Chanchamayo area (central jungle). Working paper. San Ramón, 98 p.

79.   MACEDO, M. 1993. Biological aspects of a Mesotropic cemadao in the vicinity of Guiaba, Mato Grosso.

80.   Campbell, P.; Comiskey, J.; Alonso, A.; Dallmeier, F.; Núñez, P.; Beltrán, H.; Baldeón, S.; Nauray, W.; De la Colina, R.; Acurio, L. & S. Udvardy. 2002. Modified Whittaker plots as an assessment and monitoring tool for vegetation in a lowland tropical rainforest. Environ Monit Assess, 76(1):19-41.

81.   Alonso, A. & F. Dallmeier (Eds.). 1998. Biodiversity Assessment and Long-term Monitoring of the Lower Urubamba Region in Peru: Cashiriari-3 Well Site and the Camisea and Urubamba Rivers. SI/MAB Series # 2. Smithsonian Institution/MAB Program. Washington, DC. 298 pages.

82.   Comiskey, J. A.; Campbell, J. P.; Alonso, A.; Mistry, S.; Dallmeier, F.; Núñez, P.; Beltrán, H.; Baldeón, S.; Nauray, W.; de la colina, R.; Acurio, L. & S. Udvardy. 2001. The vegetation communities of the Lower Urubamba Region, Peru. In: Alonso, A. F.

83.   Martinez Bardales, Clessy L. 2010. Structural Analysis of a Medium Terrace Forest in the Allpahuayo-Mishana National Reserve, Loreto- Peru. Thesis for the

degree of Engineer in Tropical Forest Ecology. School of Professional Formation of Engineering in Tropical Forest Ecology. Faculty of Forestry Sciences. Iquitos-Peru.

84.   KENNY SANGAMA GONZALES. Horizontal Structure and Diversity of a High Terrace Forest in a Private Property in the Amazon River Basin, District of Punchana, Loreto-Peru. Thesis for the professional degree of Engineer in Tropical Forest Ecology. Faculty of Forestry Sciences. Iquitos - Peru. 2015.

85.   AMARAL VERISSIMO, A.; BARRET, P.; VIDAL, E. 2005. Forests forever. Manual for timber production in Amazonia. WWF, AMAZON, USAID ASDI. Lima - Peru. 161 p.

86.   SAHUARICO FACHIN, ABNER ANDERSON. 2010. Study of the Natural Regeneration of a Middle Terrace Forest, Nanay River Basin. Iquitos, Loreto, Peru. Thesis for the degree of Forestry Engineer. Faculty of Forestry Sciences. Iquitos - Peru.

87.   ROJAS TUANAMA, R.; TELLO ESPINOZA, RODIL. 2006. Abundance and Stock of Natural Regeneration of Forest Species in the Varillal Forest of CIEFOR, Iquitos - Peru. Faculty of Forestry Sciences. Iquitos - Peru.

88.   FREITAS VÁSQUEZ, ROSA MARÍA. 2017. Analysis of the Natural Regeneration of two High Terrace Forests of the Community of Salvador Rio Napo, Department of Loreto - Peru. Thesis for the degree of Forestry Engineer. Faculty of Forestry Sciences. Iquitos - Peru.

89.   WADSWORTH, F. 2000, Primary forests and their productivity. In: Forest production for tropical America. Handbook of Agriculture 710-S. USDA. Washingtond, OC. Pg 69-109.

90.   ROLLET, V. 1971. Natural regeneration in dense evergreen forests of the Venezuelan Guyana Plain. Boletin del Instituto Forestal Latinoamericano de Investigación y Capacitación (Venezuela). (35) 39-73 p.

91.   WHITMORE, T. 1984. Tropical Rain forest of the Far East. Oxford. G. B. Clarendon Press. 341 p.

92.   SCHULZ, J. P. 1967. The Natural Regeneration of the Tropical Mesofrtic Rainforest of Surinam, after its utilization. Boletin del Instituto Capacitación. Venezuela (23). 27 p.

93.   SCHWYZER, A., 1981. Survey of Natural Regeneration and its use in reforestation. Jenaro Herrera Integral Rural Settlement Project. Technical Bulletin No 07. Iquitos- Peru. 18 p.

94. FINEGAN, B. 1992. Ecological Bases for Silviculture. V International Intensive Course on Silviculture and Management of Tropical Natural Forests - CA TIE- Costa Rica. 170 p.

95. HARTSHORN, A. 1980. Dynamics of Neotropical Forests. Facsimile Series No 08. Centro Científico Tropical San José de Costa Rica. Costa Rica. 26 p.

96. LOUMAN, By STANLEY, S. 2002, Analysis and interpretation of forest inventory results: In: L. Orosco and C. Brumer (eds.). Brumer (editors). Forest inventory for broadleaved forests in Central America. Technical Series, Technical Manual No. 50, Tropical Agricultural Research and Higher Education Center. CATIE. Turrialba, Costa Rica. 263 p.

97. PAIMA ESPINOZA, PIERO C. Structure and Floristic Composition of the Middle Terrace Forest Adjacent to the Arboretum "el huayo", CIEFOR- Puerto Almendras, Nanay River, Iquitos-Peru. Thesis for the degree of Engineer in Tropical Forest Ecology. Faculty of Forestry Sciences - UNAP. Iquitos - Peru. 2012. vi-vii p.

98. Pitman, N.C A. 2005. An overview of the Los watershed, Madre de Dios, southeastem Peru. ACCA. Unpublished manuscript. 51 p.

99. GARCÍA, G. J., CLAUSSI, A.; MARMILLOD, D.; and BLASER, J., 1975. Integral Study of a Tropical Rain Forest in the Jenaro Herrera Area (Iquitos).

100. Report of the National Forest and Wildlife Inventory of Peru. National Forestry and Wildlife Service. General Directorate of Forestry and Wildlife Information and Management. Directorate of Inventory and Valuation. December. 2019. 144 p.

101. Zonificación Ecológica y Económica Temático Forestal, Provincia de Alto Amazonas (2013). Informe de Evaluación del Temático Forestal. Research Institute of the Peruvian Amazon. Provincial Municipality of Alto Amazonas. Regional Government of Loreto. 21 p.

102. Peruvian Amazon Research Institute, IIAP. World Bank. 2002. Ecological Economic Zoning Study of the Nanay River Basin. Iquitos - Peru.

103. Martínez, P. 2010. Ecological and Economic Mesozonification Project for the Sustainable Development of the Apurimac River Valley-VRA. Iquitos-Peru. 64 p.

ANNEXES

Field format

Department   :
Province :
District   :
Place of collection   :
Coordinates   :
Date of collection   :
Name of the collector   :

NI	Family	Scientific Name	Name Com)n	d (cm)	cd	h	ca
1							
2							
3							
4							
5							
6							
7							
8							
9							
10							
<..							

Legend: diameter (d), diameter class (cd), height (h), altimetric class (ca).

ANNEX N° 02

Table N° 49: Floristic composition of three (3) evaluated plots.

N°	Family	Scientific Name	Common name
1	Lecythidaceae	Eschweilera coriacea (A. DC.) S. A. Mori	black machimango
2	Urticaceae	Pourouma tomentosa Mart.	sacha ubilla
3	Urticaceae	Pourouma tomentosa Mart.	sacha ubilla
4	Lauraceae	Ocotea aciphylla (Nees) Mez	cinnamon moena
5	Lauraceae	Ocotea olivacea A. C. Sm.	YELLOW SPRING
6	Fabaceae	Parkia velutina Benoist	pashaco
7	Araliaceae	Dendropanax umbellatus (Ruiz & Pav.) Decne. & Planch.	caspi phosphorus
8	Elaeocarpaceae	Sloanea guianensis (Aubl.) Benth.	casha huayo
9	Sapotaceae	Ecclinusa lanceolata (Mart. & Eichl.) Pierre	caimitillo
10	Fabaceae	Parkia velutina Benoist	pashaco
11	Araliaceae	Dendropanax umbellatus (Ruiz & Pav.) Decne. & Planch.	caspi phosphorus
12	Lauraceae	Ocotea argyrophylla Ducke	brown leaf moena
13	Fabaceae	Parkia velutina Benoist	pashaco
14	Fabaceae	Swartzia benthamiana Miq.	shimbillo steel
15	Euphorbiaceae	Hevea pauciflora (Spruce ex Benth.) Müll. Arg.	shiringa
16	Araliaceae	Dendropanax umbellatus (Ruiz & Pav.) Decne. & Planch.	caspi phosphorus
17	Fabaceae	Parkia velutina Benoist	pashaco
18	Sapotaceae	Ecclinusa lanceolata (Mart. & Eichl.) Pierre	caimitillo
19	Sapotaceae	Ecclinusa lanceolata (Mart. & Eichl.) Pierre	caimitillo
20	Fabaceae	Parkia velutina Benoist	pashaco
21	Elaeocarpaceae	Sloanea guianensis (Aubl.) Benth.	casha huayo
22	Myristicaceae	Iryanthera juruensis Warb.	red cumalilla
23	Sapotaceae	Ecclinusa lanceolata (Mart. & Eichl.) Pierre	caimitillo
24	Fabaceae	Inga tessmannii Harms	shimbillo
25	Euphorbiaceae	Hevea pauciflora (Spruce ex Benth.) Müll. Arg.	shiringa
26	Sapotaceae	Ecclinusa lanceolata (Mart. & Eichl.) Pierre	caimitillo
27	Elaeocarpaceae	Sloanea durissima Spruce	cepanchina
28	Lauraceae	Ocotea oblonga (Meisn.) Mez.	shicshi moena
29	Fabaceae	Parkia velutina Benoist	pashaco
30	Elaeocarpaceae	Sloanea guianensis (Aubl.) Benth.	casha huayo
31	Fabaceae	Hymenolobium nitidum Benth	mari mari
32	Euphorbiaceae	Hevea pauciflora (Spruce ex Benth.) Müll. Arg.	shiringa
33	Elaeocarpaceae	Sloanea guianensis (Aubl.) Benth.	casha huayo
34	Lauraceae	Ocotea gracilis (Meisn.) Mez.	odorless moena
35	Sapotaceae	Ecclinusa lanceolata (Mart. & Eichl.) Pierre	caimitillo
36	Lauraceae	Ocotea gracilis (Meisn.) Mez.	odorless moena
37	Sapotaceae	Ecclinusa lanceolata (Mart. & Eichl.) Pierre	caimitillo
38	Elaeocarpaceae	Sloanea guianensis (Aubl.) Benth.	casha huayo
39	Fabaceae	Parkia velutina Benoist	pashaco
40	Burseraceae	Protium ferrugineum (Engl.) Engl.	red copal

41 Lauraceae	Anaueria brasiliensis Kosterm.	añuje rumo
42 Lauraceae	Ocotea myriantha (Meisn.) Mez.	heron moena
43 Moraceae	Ficus guianensis Desv.	renaco
44 Burseraceae	Protium altsonii Sandwith	copal
45 Euphorbiaceae	Hyeronima oblonga (Tul.) Müll. Arg.	high altitude caspi mouse
46 Lecythidaceae	Eschweilera coriacea (A. DC.) S. A. Mori	white machimango
47 Lecythidaceae	Eschweilera parvifolia Mart. ex A. DC.	machimango
48 Lauraceae	Beilschmiedia sp.	cinnamon moena
49 Fabaceae	Swartzia klugii (R.S. Cowan) Torke	sacha cumaceba
50 Lauraceae	Ocotea bracteosa (Meisn.) Mez.	cinnamon moena
51 Meliaceae	Guarea macrophylla Vahl	requia
52 Melastomataceae	Miconia symplectocaulos Pilg.	caracha caspi
53 Urticaceae	Cecropia latiloba Miq.	white cetico
54 Fabaceae	Parkia velutina Benoist	pashaco
55 Lauraceae	Ocotea myriantha (Meisn.) Mez.	heron moena
56 Annonaceae	Xylopia cuspidata Diels	caspi turtle
57 Araliaceae	Dendropanax umbellatus (Ruiz & Pav.) Decne. & Planch.	caspi phosphorus
58 Fabaceae	Parkia velutina Benoist	pashaco
59 Sapotaceae	Ecclinusa lanceolata (Mart. & Eichl.) Pierre	caimitillo
60 Lauraceae	Nectandra paucinervia Coe-Teix.	moena
61 Lauraceae	Nectandra paucinervia Coe-Teix.	moena
62 Fabaceae	Parkia velutina Benoist	pashaco
63 Fabaceae	Macrolobium microcalyx Ducke	saint caspi
64 Lauraceae	Ocotea myriantha (Meisn.) Mez.	heron moena
65 Elaeocarpaceae	Sloanea guianensis (Aubl.) Benth.	casha huayo
66 Lauraceae	Ocotea myriantha (Meisn.) Mez.	heron moena
67 Elaeocarpaceae	Sloanea guianensis (Aubl.) Benth.	casha huayo
68 Lauraceae	Ocotea myriantha (Meisn.) Mez.	heron moena
69 Annonaceae	Xylopia cuspidata Diels	caspi turtle
70 Sapotaceae	Micropholis venulosa (Mart. & Eichl.) Pierre	ballet
71 Anacardiaceae	Tapirira guianensis Aubl.	wira caspi
72 Euphorbiaceae	Hevea pauciflora (Spruce ex Benth.) Müll. Arg.	shiringa
73 Urticaceae	Pourouma mollis Trécul	sacha ubilla
74 Myristicaceae	Iryanthera paraensis Huber	cumalilla
75 Humiriaceae	Sacoglottis amazonica Mart.	shungo parrot
76 Lecythidaceae	Eschweilera albiflora (A. DC.) Miers	machimango
77 Urticaceae	Pourouma tomentosa Mart.	sacha ubilla
78 Lecythidaceae	Eschweilera tessmannii Knuth	machimango colorado
79 Sapotaceae	Chrysophyllum bombycinum T. D. Penn.	balata large sheet
80 Annonaceae	Xylopia cuspidata Diels	caspi turtle
81 Annonaceae	Xylopia cuspidata Diels	caspi turtle
82 Fabaceae	Inga brachyrhachis Harms	shimbillo
83 Sapotaceae	Micropholis egensis (A. DC.) Pierre	quinilla
84 Sapotaceae	Micropholis venulosa (Mart. & Eichl.) Pierre	ballet
85 Malpighiaceae	Byrsonima stipulina J. F. Macbr.	flag of caspi

	Family	Species	Common name
86	Fabaceae	Parkia velutina Benoist	pashaco
87	Euphorbiaceae	Hevea pauciflora (Spruce ex Benth.) Müll. Arg.	shiringa
88	Annonaceae	Xylopia cuspidata Diels	caspi turtle
89	Fabaceae	Parkia velutina Benoist	pashaco
90	Lauraceae	Ocotea gracilis (Meisn.) Mez.	odorless moena
91	Fabaceae	Parkia velutina Benoist	pashaco
92	Fabaceae	Macrolobium microcalyx Ducke	saint caspi
93	Apocynaceae	Macoubea guianensis Aubl.	huayo syrup
94	Sapotaceae	Micropholis venulosa (Mart. & Eichl.) Pierre	ballet
95	Lauraceae	Ocotea gracilis (Meisn.) Mez.	odorless moena
96	Myristicaceae	Iryanthera juruensis Warb.	red cumalilla
97	Anacardiaceae	Tapirira guianensis Aubl.	wira caspi
98	Lauraceae	Ocotea gracilis (Meisn.) Mez.	odorless moena
99	Fabaceae	Macrolobium microcalyx Ducke	saint caspi
100	Myrtaceae	Calyptranthes ruiziana O. Berg	guayabilla
101	Euphorbiaceae	Hevea pauciflora (Spruce ex Benth.) Müll. Arg.	shiringa
102	Fabaceae	Parkia velutina Benoist	pashaco
103	Myrtaceae	Eugenia florida DC.	sacha guava
104	Fabaceae	Parkia velutina Benoist	pashaco
105	Annonaceae	Xylopia cuspidata Diels	caspi turtle
106	Annonaceae	Xylopia cuspidata Diels	caspi turtle
107	Annonaceae	Xylopia cuspidata Diels	caspi turtle
108	Annonaceae	Xylopia cuspidata Diels	caspi turtle
109	Malpighiaceae	Byrsonima stipulina J. F. Macbr.	flag of caspi
110	Myristicaceae	Iryanthera juruensis Warb.	red cumalilla
111	Euphorbiaceae	Hevea pauciflora (Spruce ex Benth.) Müll. Arg.	shiringa
112	Olacaceae	Tetrastylidium peruvianum Sleumer	yutubanco
113	Myristicaceae	Iryanthera tricornis Ducke	pucuna caspi
114	Clusiaceae	Moronobea coccinea Aubl.	sulfur caspi
115	Fabaceae	Tachigali tessmannii Harms	tangarana
116	Fabaceae	Parkia velutina Benoist	pashaco
117	Burseraceae	Protium ferrugineum (Engl.) Engl.	red copal
118	Lecythidaceae	Eschweilera coriacea (A. DC.) S. A. Mori	white machimango
119	Sapotaceae	Pouteria torta (Mart.) Radlk.	white quinilla
120	Meliaceae	Guarea macrophylla Vahl	requia
121	Lauraceae	Aniba perutilis Hemsl.	YELLOW SPRING
122	Myristicaceae	Iryanthera polyneura Ducke	cumala colorada
123	Euphorbiaceae	Hevea pauciflora (Spruce ex Benth.) Müll. Arg.	shiringa
124	Apocynaceae	Aspidosperma schultesii Woodson	quillo bordón
125	Euphorbiaceae	Alchornea triplinervia (Spreng.) Müll. Arg.	Caspian mosquito
126	Euphorbiaceae	Alchornea triplinervia (Spreng.) Müll. Arg.	Caspian mosquito
127	Euphorbiaceae	Alchornea triplinervia (Spreng.) Müll. Arg.	Caspian mosquito
128	Sapotaceae	Pouteria torta (Mart.) Radlk.	white quinilla
129	Euphorbiaceae	Alchornea triplinervia (Spreng.) Müll. Arg.	Caspian mosquito
130	Myristicaceae	Osteophloeum platyspermum (A. DC.) Warb.	weeping cumala
131	Myristicaceae	Iryanthera lancifolia Ducke	cumala colorada

132 Urticaceae	Pourouma minor Benoist	sacha ubilla
133 Euphorbiaceae	Alchornea triplinervia (Spreng.) Müll. Arg.	Caspian mosquito
134 Arecaceae	Socratea exorrhiza (Mart.) H. Wendl.	casha pona
135 Euphorbiaceae	Alchornea triplinervia (Spreng.) Müll. Arg.	Caspian mosquito
136 Lauraceae	Ocotea gracilis (Meisn.) Mez.	odorless moena
137 Fabaceae	Swartzia benthamiana Miq.	shimbillo steel
138 Fabaceae	Swartzia benthamiana Miq.	shimbillo steel
139 Fabaceae	Swartzia benthamiana Miq.	shimbillo steel
140 Annonaceae	Xylopia cuspidata Diels	caspi turtle
141 Metteniusaceae	Dendrobangia boliviana Rusby	avocado caspi
142 Burseraceae	Protium ferrugineum (Engl.) Engl.	red copal
143 Burseraceae	Crepidospermum prancei Daly	white copal
144 Metteniusaceae	Dendrobangia boliviana Rusby	avocado caspi
145 Metteniusaceae	Dendrobangia boliviana Rusby	avocado caspi
146 Sapotaceae	Micropholis guyanensis (A. DC.) Pierre	balata saponin
147 Sapotaceae	Pouteria cladantha Sandwith	quinilla
148 Fabaceae	Parkia velutina Benoist	pashaco
149 Euphorbiaceae	Nealchornea yapurensis Huber	caspi
150 Myristicaceae	Iryanthera juruensis Warb.	red cumalilla
151 Sapotaceae	Ecclinusa lanceolata (Mart. & Eichl.) Pierre	caimitillo
152 Sapotaceae	Ecclinusa lanceolata (Mart. & Eichl.) Pierre	caimitillo

Printed by Books on Demand GmbH, Norderstedt / Germany